AF590915

Electronic Measurement Techniques

Andrew D. Balmos

Sutton R. Hathorn

Brooke A. Parks, Editor

Copyright ©2019 Andrew D. Balmos and Sutton R. Hathorn.

All rights reserved. No part of this book may be reproduced in any form or by any electronic or mechanical means, including information storage and retrieval systems, without permission in writing from the publisher.

Neither the author nor the publisher assumes any responsibility or liability whatsoever on behalf of the consumer or reader of this material. The resources in this book are provided for informational purposes only and should not be used to replace the judgment of an experienced engineer. Neither the author nor the publisher can be held responsible for the use of the information provided within this book. Always take proper safety precautions before using any electrical equipment.

ISBN: 978-1-7334389-1-9

Published by Balmos Hathorn

Acknowledgments

No project of this magnitude is possible without many hands. We are particularly grateful to Matthew Swabey, Barrett Robinson, Yang Wang, and Zachary Vander Missen for their content contributions. Additionally, this book would not be the classroom-ready text that it is without the brave graduate teaching assistants who taught from and improved on our early editions. Many minds have enriched this project with their ideas, and we are truly grateful.

Finally, we thank our families for their understanding and patience during this adventure. Without their support we would not have succeeded in our endeavors.

Online Resources

We have set up an online cache of resources that students and instructors may find useful. The resources are available at `http://electronicdevicesanddesign.com`.

An online circuit simulator has been provided with pre-made schematics of some of the experiments. We encourage you to experiment with the online simulator to help visualize how the circuits are functioning.

For the in-lab simulations, we recommend the use of LTSpice. Links to the software and tutorials are available online. Additionally, we have added models of devices not included with the LTSpice software that are used in the experiments.

The website also contains an errata for any errors found after publication and links to the datasheets referred to by the text.

Contents

EXPERIMENT 1

Introduction Experiment

1.1 Application

Many types of equipment are used to measure and test electrical circuits in the laboratory. We use this equipment to measure how circuits are behaving and to aid with designing and debugging systems. Developing a thorough understanding of measurement equipment is critical to circuit development. To the dismay of many new students, even expensive test equipment is not ideal and cannot always be trusted. Understanding limitations and equivalent circuit models of all your tools will help you avoid mistakes and misinterpretations of results.

Much of the equipment you will learn how to use in this experiment has been present in electrical engineering labs for many years, but newer equipment continues to push the accuracy and precision of measurement possible. Nonetheless, the core concepts involved in measurement are the same as in old equipment like the multimeter pictured in figure 1.1. A good understanding of how measurement equipment works will enable you to use any similar piece of equipment, old or new.

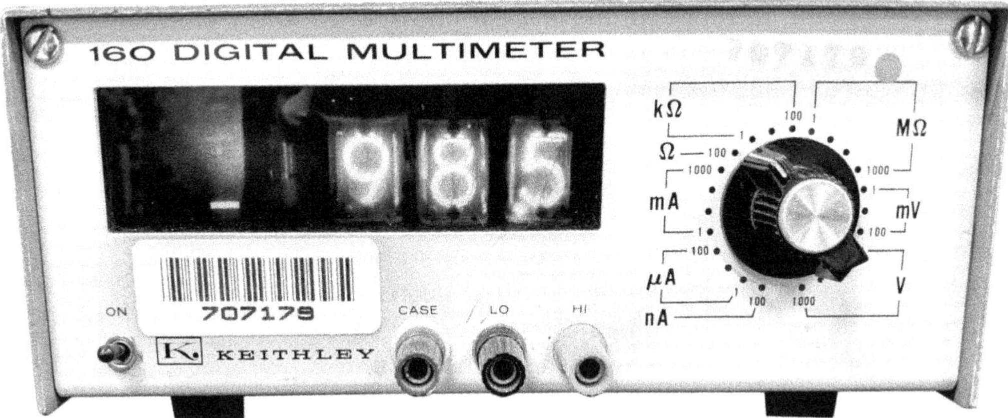

Figure 1.1: A Keithley 160 Digital Multimeter with a nixie tube display. It was originally designed and built in the 1960s and is still relatively accurate today!

1.2 Current

Current is the fundamental physical reaction of a completed electric circuit. Current cannot flow unless a completely closed circuit is formed. The flow of electrons creates heat in devices and generates the electromagnetic fields that power motors, radio broadcasts, and more. While many modern design practices focus on the manipulation of *voltage*, current is a critical detail that is often too easily forgotten by new engineers; however, working with current control can solve a number of unique problems.

1.2.1 So, what is current?

Current is a measure of charge (typically electrons) moving through a given area. Typically, this 'area' is a wire. The SI unit for current is the ampere (A), defined as one coulomb of charge through a surface in one second. Mathematically, current is defined as

$$I = \frac{Q}{t} \tag{1.1}$$

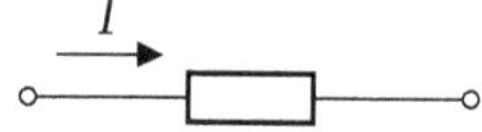

Figure 1.2: Current flow into a device represented here as a rectangle.

where I is current in amperes, Q is the amount of charge in coulombs, and t is time in seconds.

Circuit diagrams indicate the flow of current by using an arrow pointing in the direction of positive current, as shown in figure 1.2.

In almost all electrical circuits, current is the result of the movement of electrons. Electrons are negatively charged, so the direction of "positive" current is actually the opposite of the direction that electrons are moving.

1.2.2 Kirchhoff's current law

Kirchhoff's current law, often abbreviated KCL, states a fundamental property of current flowing in and out of objects. KCL states that all of the current entering an object must be equal to all of the current leaving an object. Mathematically, KCL can be expressed by equation (1.2).

$$\sum I_{\text{in}} = \sum I_{\text{out}} \tag{1.2}$$

Kirchhoff's current law allows us to solve for an unknown current in a circuit given that we know the other currents flowing through a circuit. For example, we can apply KCL to the connection of the three wires in figure 1.3. The point where wires connect is called a "node." At this node, all of the current entering must be equal to the current leaving. Therefore, by KCL,

$$I_1 + I_2 = I_3.$$

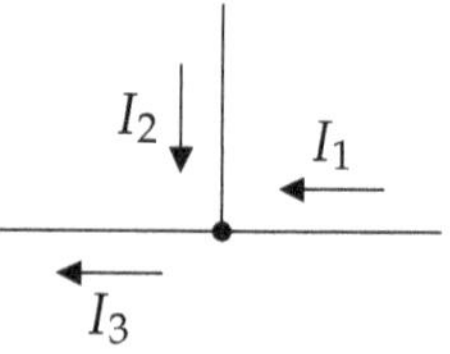

Figure 1.3: KCL example.

1.3 Voltage

Voltage describes the available electric potential. While somewhat of a misnomer, it may help to initially think of it as the "pressure" with which current is pushed. That said, voltage is present even when a circuit is not complete, as is water pressure when no water is flowing from a spigot. In many ways voltage is analogous to potential energy; it describes how much potential there is for current flow.

Voltage can also be seen as a way to measure electrical energy that is lost due to the movement of charge. That is, when voltage drops from one side of a circuit element to the another, the difference is the quantity of potential consumed by that device. These concepts will clarify as you continue your study of electronics.

1.3.1 So, what is voltage?

More precisely, voltage is the difference in electrical potential energy between *two* points. The SI unit for voltage is the volt (V), defined as one joule of work per coulomb of charge moved between the voltage's reference points. For example, a voltage of 1 V means that one coulomb of charge loses 1 joule of potential energy when moving from the starting (positive) measurement point to the ending (negative) measurement point.

The voltage between two points is written as V_{AB}, where point A is positive and point B is negative. V_{AB} is how much higher the potential energy is at point A than point B. If a voltage only has one subscript, such as V_A, then that indicates that the negative point is ground. Ground is typically used as a common reference point for voltages in a circuit and indicated by the symbols like those in figure 1.4.

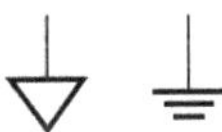

Figure 1.4: Ground Symbols.

It is important to know which ground a voltage is measured with respect to. A circuit may occasionally have different ground references that should be marked with unique ground symbols. Always be clear when labeling circuits with multiple ground points, even if the extra labeling seems tedious.

Voltage is indicated on circuit diagrams by using both a "+" and "-" sign at the positive and negative measurement points, respectively. It may also be drawn as a voltage label at a node, which means that ground is used as the negative measurement reference. An example of the two methods is shown in figure 1.5. Because of the location of ground, $V_{CB} = V_C$ and $V_{AB} = V_A$.

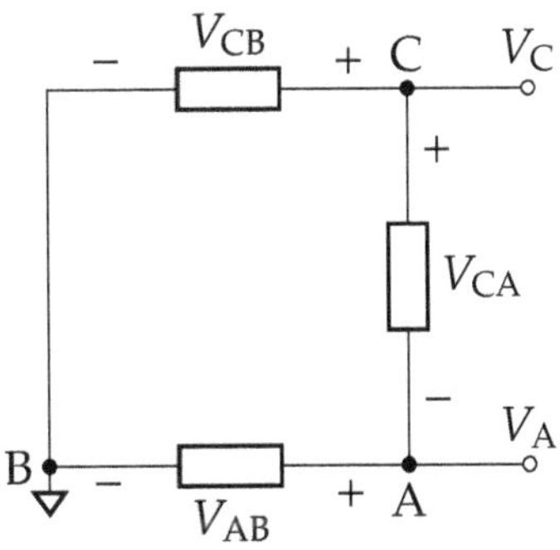

Figure 1.5: Example voltage labels.

1.3.2 Kirchhoff's voltage law

Kirchhoff's voltage law, abbreviated KVL, states that the sum of voltages across a loop must be equal to zero. That is, the total change in potential energy in a closed path is zero. This can be thought of like altitude: if you travel in a closed path, the altitude that you start at will be the altitude you finish at no matter what path you take. Mathematically, KVL can be described by equation (1.3).

$$\sum_{\text{loop}} V = 0 \tag{1.3}$$

We can apply KVL to the circuit in figure 1.5 around the closed path from B to A to C and back to B. The resulting equation is given in equation (1.4). Note that V_{CB} is subtracted because it indicates the increase in potential from point B to C, but we are writing the KVL equation for the path going from point C to B.

$$V_{\text{AB}} + V_{\text{CA}} - V_{\text{CB}} = 0 \tag{1.4}$$

1.4 Power

Power is the rate of energy transfer. It is measured in joules per second. The SI unit for power is the Watt (W). Power can be computed as

$$P = \frac{\Delta E}{\Delta t}$$

where ΔE is the energy transfered in joules and Δt is length of the transfer event in seconds.

Power is similar to work in that it measures the amount of change in the environment, except that it also accounts for the length of time over which the change occurred. For example, carefully slowing from a high speed to stationary when approaching a red light requires the same amount of work to be done by the car as stopping abruptly to avoid an accident; however, stopping to avoid the accident requires significantly more power due to the reduced time window.

1.4.1 Electric power

Applying the aforementioned definition of current and voltage, one finds that that electric power can be written in terms of voltage and current by

$$P = \frac{E}{t} = \frac{Q \cdot V}{t} = I \cdot V \tag{1.5}$$

where E is the number of joules moved by the current and Q is the number of coulombs moved through the voltage V. Finally, equation (1.1) is used to rewrite the formula in terms of only voltage and current. In circuits, power is often transferred from an energy source to heat, motion, or light.

1.4.2 Passive sign convention

The standard for labeling passive components in circuit diagrams is the passive sign convention shown in figure 1.6. It dictates that current flows into the positive voltage on a circuit element. When following the passive sign convention, positive voltage and positive current will result in positive power. When components dissipate power, the voltage and current will have the same sign therefore making analysis easier.

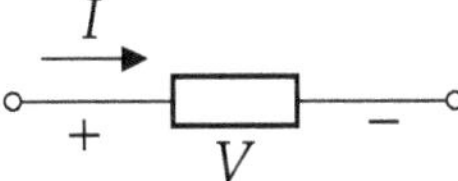

Figure 1.6: Element labeled with passive sign convention.

1.5 The resistor

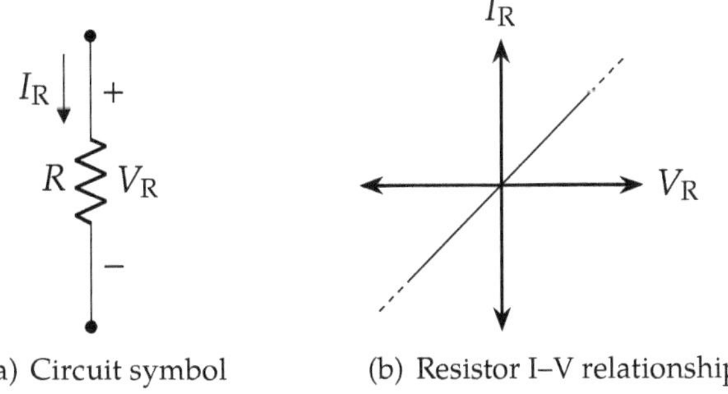

(a) Circuit symbol (b) Resistor I–V relationship

Figure 1.7: The resistor.

A resistor is a component that tends to "resist" the flow of current through itself. Resistance is the constant of proportionality between voltage drop over and current flow through a resistor. The SI unit for resistance is the Ohm (Ω), and is defined as 1 volt drop over a 1 Ohm resistor results in a 1 amp current. A common symbol for a resistor is shown in figure 1.7(a), annotated with voltage drop and current flow.

1.5.1 Ohm's law

Ohm's law states that the voltage of a conductor is proportional to the current where the constant of proportionality is the resistance. An example plot of Ohm's law is shown in figure 1.7(b). This can be stated mathematically as the formula

$$V = I \cdot R \tag{1.6}$$

Older notation uses E in place of V, therefore sometimes $E = I \cdot R$.

where R is the resistance in Ohms, V is the voltage across the resistor and I is the resistor's current in amperes.

Occasionally, Ohm's law is written in terms of conductance instead of resistance, where the conductance G, measured in Siemens, is simply the inverse of resistance.

$$I = G \cdot V$$

It is important to note that these results are only true for ideal resistors, but we can often approximate real circuits as ideal in order to take advantage of the simplicity of Ohm's law.

When a resistor's environment changes, the resistor itself may change. For example, as a resistor heats from electrical work, the component's chemical properties can be altered and it becomes a new resistor with a new resistance.

1.5.2 Power dissipated in a resistor

Plugging Ohm's law, equation (1.6), into the definition of DC electric power, equation (1.5), results in the following useful formulas:

$$P = V \cdot I = I^2 R = \frac{V^2}{R} \tag{1.7}$$

1.5.3 The physical resistor

Resistors can take on many shapes and sizes. The most common type of axial resistor used in circuits is made from a carbon or metal film. The resistor will have a set of colored stripes on it to indicate the value and tolerance of the resistor. A guide for reading the color bands is in appendix C on page 220.

Figure 1.8: Various types of resistors.

1.6 Everything is a circuit component

Keep in mind that all physical things, whether designed to be used in a circuit or not, have some relationship between the voltage they *drop*(1) and the current *flowing* through them. Even a rock or a chunk of plastic will alter, albeit slightly, the operation of a circuit. The relation may not be that useful to a designer; nevertheless, if connected it *will* alter the circuit.

(1) The term "dropped" is common jargon used to describe the voltage difference between two leads of a component.

These voltage-current relationships are called I–V curves or characteristics. In practice the curves are often non-linear functions of voltage[(2)] and can be rather hard to work with mathematically. Fear not, as the linear characteristic of the resistor will be comparatively simple to analyze and can model a wide variety of devices.

[(2)] Most semiconductor technology is made up of highly non-linear devices called transistors.

It is surprising how far one can get by *carefully* applying linear models for non-linear circuits by approximating non-linear components as simple resistors and sources. Even though the circuit is not made from the exact components in the model, they will behave almost exactly as if that was the case. We will use linear models to understand the impact a piece of measurement equipment will have on a circuit.

1.7 DC measurement equipment

Several different types of equipment are used for measurement of circuits. Power can be supplied to a circuit using a voltage or current supply and the voltages and currents are measured with a voltmeter or ammeter, respectively. Resistance can be measured using an ohmmeter. Instead of having several pieces of equipment for each type of meter, a multimeter is a device that combines a voltmeter, ammeter, and ohmmeter into one.

1.7.1 Ideal ammeter and voltmeter

The ideal ammeter and voltmeter make up the tools needed for measuring various quantities in a circuit. An ideal ammeter will measure the current flowing through it and relay the information to the user in an understandable way. The symbol for the ideal ammeter is a circle with an A in it, as shown in figure 1.9(a). The quantity measured by the meter is all of the current flowing through said meter, and the ideal meter has a voltage of 0 V across it. Because the voltage is 0, there is no power dissipated in the meter. When connecting an ammeter to a circuit, it should be treated as a short circuit.

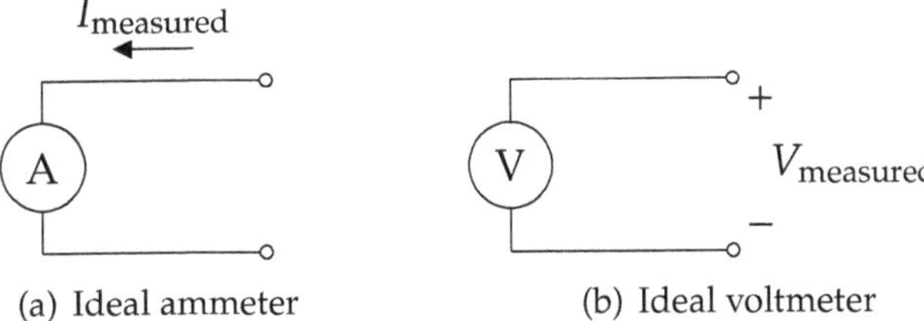

(a) Ideal ammeter (b) Ideal voltmeter

Figure 1.9: Meter types.

The ideal voltmeter is drawn as a circle with a V in it, as shown in figure 1.9(b). It measures the potential difference between two points in a circuit. Ideal voltmeters have a current of 0 A. When connecting a voltmeter to a circuit, it can be treated as an open circuit.

Example 1.7.1: Making a current measurement

This example will walk through the procedure for measuring I_3 in figure 1.10.

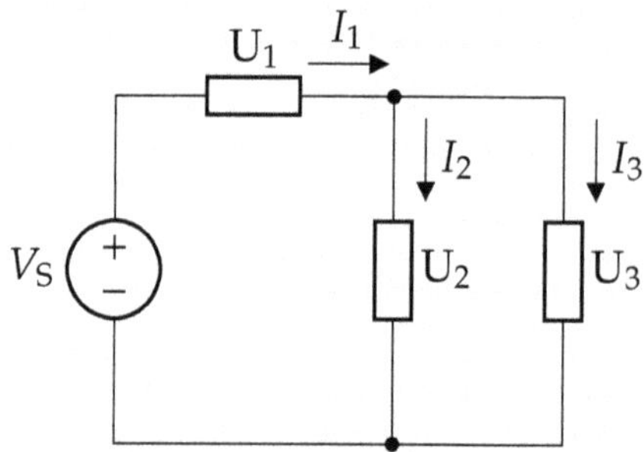

Figure 1.10: An electrical circuit

In order to measure current, an ammeter needs to be inserted into the path of the current. For this circuit, I_3 can be measured by disconnecting either half of U_3 and adding the ammeter in series.

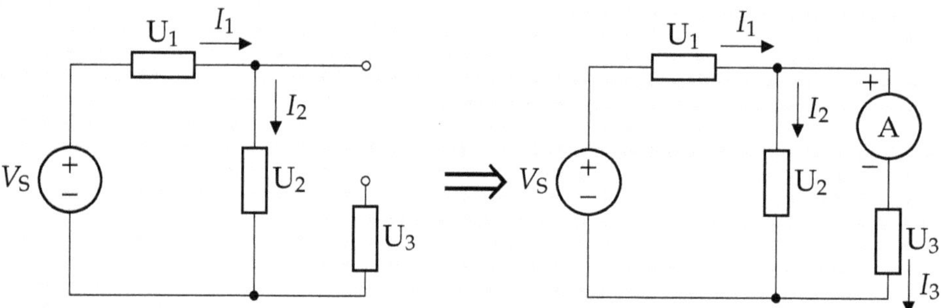

Figure 1.11: Adding the ammeter

Figure 1.11 shows the procedure for adding the ammeter. First, the top node of U_3 is disconnected, and then the ammeter is inserted in series. The ammeter measures current flowing through the positive node to the negative node. The positive node is typically a red cable, and the negative is typically a black cable.

Due to KCL, all of the current flowing through the ammeter much also flow through U_3, so the current shown on the ammeter is the measurement we want.

1.7.2 Ideal voltage and current sources

Circuits need to be energized by a voltage or current source in order to do something useful. The ideal voltage source is drawn as a circle with a plus and minus sign as shown in figure 1.12(a). The symbol indicates that there is a device that generates a potential difference between its terminals equal to the labeled voltage, where the plus terminal is at a higher potential. Other common symbols for voltage sources are shown in figures 1.13 and 1.14. The ideal voltage source will source (send out) and sink (take in) current in order to ensure that the potential difference between the terminals is always what is labeled.

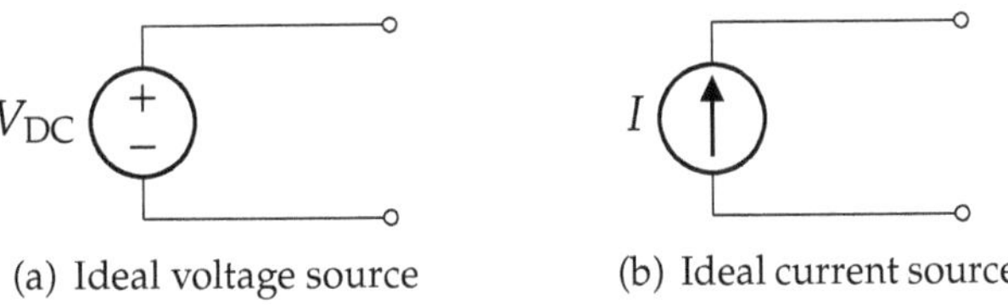

(a) Ideal voltage source (b) Ideal current source

Figure 1.12: Source types

Current sources are drawn as a circle with an arrow pointing in the direction of current flow as shown in figure 1.12(b). An ideal current source will always have the labeled current flowing through it and can have any voltage across it. In practice, current sources are not very common; however, they are often used in modeling more complicated circuits.

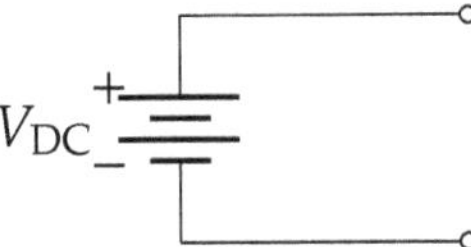

Figure 1.13: Battery voltage source

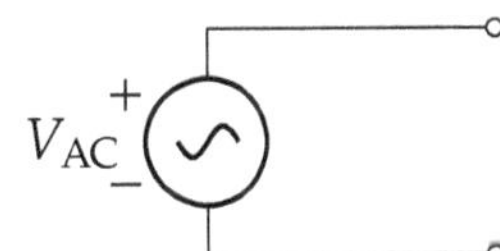

Figure 1.14: Time-varying source

1.8 Multimeters

Voltmeters and ammeters are typically packaged together in a single device called a multimeter. Multimeters are devices that can measure more than one electrical quantity. Nearly all multimeters are capable of measuring voltage, current, and resistance. Some meters have additional features such as frequency measurement or capacitance measurement.

Current measurement Current is measured by wiring an ammeter in series at the point of interest in the circuit. The circuit must be powered by an external supply at the time of the test because the ammeter does not provide any energy.

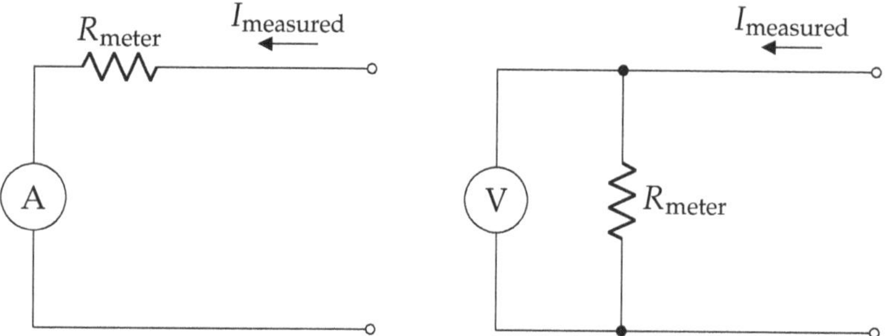

Figure 1.15: Equivalent circuit model of a real ammeter. The left models equipment that directly measures the test circuit current. The right models equipment that measures a voltage drop caused by the test current. In practice R_{meter} is small.

There are two primary ammeter models, as seen in figure 1.15. The first directly measures the test circuit current flowing through the ammeter. Alternatively, a voltage induced by the circuit's current flowing through R_{meter} is measured. In this case, Ohm's law is used to compute the current.

Direct current measurement is most common in older analog meters. Most digital meters measure the voltage across a known resistance.

From the device under test's (DUT's) perspective, the two models are equivalent. In effect, an ammeter appears as a very small resistance. Therefore, when the meter is connected in series with the test circuit, it *should* have little impact on the measurements and waste insignificant power. One still must verify that the meter's resistance is in fact small enough to not alter the circuit under test.

Voltage measurement Voltage is measured by wiring the voltmeter in parallel across the points of interest in the circuit. The circuit must be powered by an external supply at the time of the test.

There are two primary voltmeter architectures, as seen in figure 1.16. The first, common in analog meters, causes the measured voltage to induce a current in the meter's internal resistance. By sensing this current and applying Ohm's law, the applied voltage can be computed as $V_{meas} = I_{meas} \cdot R_{meter}$. Alternatively, the test voltage is directly measured.(3)

(3) For example, an analog to digital converter (ADC) may be used to directly digitize the test voltage in a digital meter.

From the device under test's perspective, the two models are equivalent. In effect, a voltmeter appears as a large resistance. Therefore, when the meter is connected in parallel to the test circuit, it *should* have little impact on the measurements and waste insignificant power. One still must verify that the meter's resistance is in fact large enough to not alter the circuit under test.

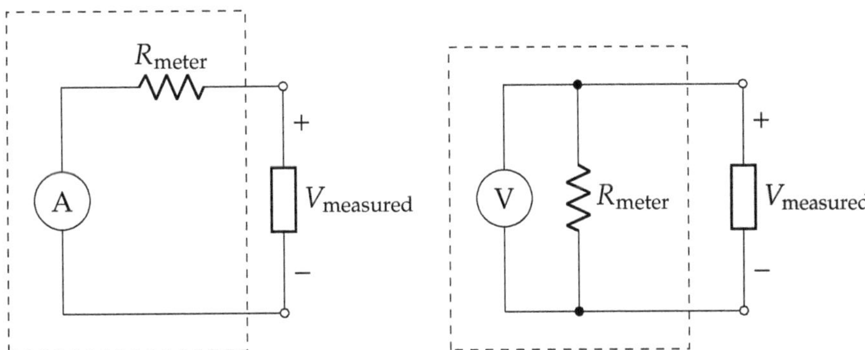

Figure 1.16: Equivalent circuit model of a real voltmeter. The left models equipment that measures current induced in the meter's resistance by the test voltage. The right circuit models equipment that directly measures the test voltage. In practice R_{meter} is large.

Resistance Resistance is measured by placing the two test leads of an ohmmeter across the nodes of interest. The component or sub-circuit must be removed from the overall circuit to make an accurate measurement. Resistance cannot be accurately measured while a component is in circuit (powered or not). This is because the rest of the circuit will either interfere with equipment's applied signal or the equivalent resistance of the entire circuit, rather than just the single resistor, will be measured.

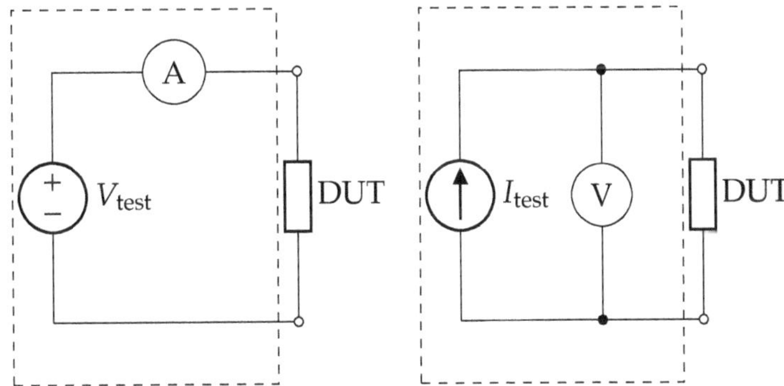

Figure 1.17: Equivalent circuit model of an ohmmeter. The left circuit models equipment that applies voltage and measures current. The right circuit models equipment that applies current and measures voltage.

There are two primary ohmmeter architectures, as seen in figure 1.17. The first, common in analog meters, applies a voltage to the circuit under test from a built-in DC supply and measures the resulting current. Using Ohm's law, the meter converts the current to a measured resistance. The alternative method, more common in digital multimeters, forces a known current through the circuit under test and measures the resulting voltage.

From the DUT's perspective, the two models are equivalent. In effect, an ohmmeter appears as a large resistance to the load. Therefore, when the meter is connected in parallel to the test circuit it *should* have little impact on the measurements and waste insignificant power.

Continuity test Many multimeters have a continuity test mode. The continuity test is used to check for the presence of a short circuit. Often times the meter will buzz when there is a short between the two leads. This feature can be used to quickly check that a circuit is correctly connected.

Figure 1.18: Analog VOM

1.8.1 Digital multimeter (DMM) versus analog volt ohm meter (VOM)

A volt ohm meter (VOM) is an analog instrument that uses a galvanometer to make current measurements. A galvanometer is an electromechanical system that converts passing current to the mechanical movement of a needle across a printed scale. As shown previously, one can measure resistance, voltage, and current with only an ammeter. Typically a VOM will have resistance, voltage, and current modes. Measurements can be read off in the appropriate units, despite the fact that the meter is actually measuring current. The technician is also required to manually select the correct measurement range. The analog scale and needle are error prone and can be imprecise if used incorrectly. Because of its limitations, analog meters find limited use in the modern lab.

Figure 1.19: Benchtop DMM

A digital multimeter (DMM) is a digital instrument that uses an analog to digital converter (ADC) to directly digitize the sensed voltage. As shown above, one can measure resistance, voltage, and current with only a voltmeter. Typically a DMM will provide various menu options to change the type of meter. After the instrument processes the internal voltage measurement, the result is shown on the screen in the correct units. Many DMMs have auto ranging functionality. The digital readout and high ADC accuracy enable high precision at relatively low cost.

Figure 1.20: Handheld DMM

While the DMM is generally technically superior to the analog VOM, the VOM still has some minor advantages. The VOM can be used for visualizing slowly varying signals. That said, an oscilloscope or root-mean-square (RMS) measurement is often a better choice in that situation.[4] Finally, as we will see in the next experiment, benchtop DMMs can be externally programmed and automated, which makes them incredibly useful for testing circuits.

[4] More to come on oscilloscopes and RMS in future experiments.

1.9 Voltage supplies

Every circuit requires some sort of voltage (or current) supply to function. Most of the time a mix of DC and AC voltages are required to complete a task. However, for now, we will restrict ourselves to just DC supplies.(5)

(5) AC power supplies (as opposed to AC low-power signals from a function generator) are often sourced using the building's wall power and a transformer or variac.

DC supplies can come in a multitude of forms, all the way from potato batteries to lab-grade test equipment.(6) The power supply most people are familiar with is a USB charger that converts an AC voltage source to a fixed DC voltage. We will spend the majority of our time with lab-grade DC supplies.

(6) At least one of the authors has made perfectly good use of a potato battery before.

Figure 1.21 is an *equivalent* circuit model of a typical DC supply. Keep in mind that, while the supply is likely a complex circuit, when operating correctly the output characteristics look a lot like an ideal voltage source and a resistor. In other words, the I–V *model* of a power supply is linear.

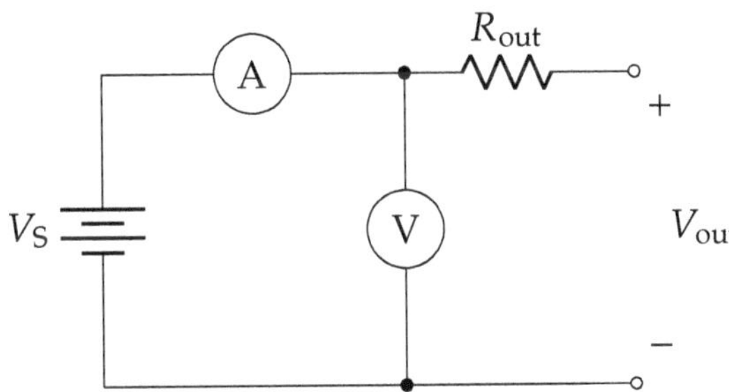

Figure 1.21: Equivalent circuit model for a DC supply.

As shown in figure 1.21, DC supplies can contain both a voltmeter and ammeter. While technically not required(7) in a DC supply, they are very useful in the laboratory and are often present in lab-grade DC supplies.

(7) The voltmeter could be replaced with an open and the ammeter replaced with a short.

Voltmeter

Usually the internal DC supply is adjustable by the user over some range. The voltmeter senses the output of this variable supply and, among other things, displays it to the user so they can accurately power their circuits.

Ammeter

In many cases the DC supply can sense the current being delivered to a load and decrease current if a preset limit is exceeded. If current limiting is available, one should get in the habit of setting it to a reasonable value before connecting the supply to a circuit.(8) Always start small with current limits.

(8) This is a feature that saves an unimaginable number of components in student laboratories each and every semester.

The limit can always be increased later, but setting the limit too high could cause a circuit to be damaged permanently.

Lab-grade supplies may allow the operator to set a current limit with the intention of it being reached. In this case the DC supply acts like a current source. This is achieved by adjusting the output voltage until the current is approximately at its limit.

Effect of R_{out}

You may be wondering what the effect of the R_{out} is on the performance of the supply. In short, it limits the amount of power the supply can provide, and preferably, it is as small as possible. Thankfully, most lab-grade supplies use some sort of internal feedback that helps reduce R_{out}. For a more complete answer, we analyze the equivalent circuit with a load in example 1.9.1.

Power envelope

The actual power being delivered varies with the load. Every DC supply has a so-called "power envelope" which defines the region of operations that are possible. An example power envelope for an ideal DC supply is shown in figure 1.22. The power supply will operate along the dashed line based on the set current and voltage limit. A lab-grade DC supply will have a fairly rectangular power envelope like the one in figure 1.22. This allows the DC supply to be used as a voltage source or current source. Cheaper power supplies may not be consistent under high current operation, resulting in the voltage droping unexpectedly or entire DC supply failure.

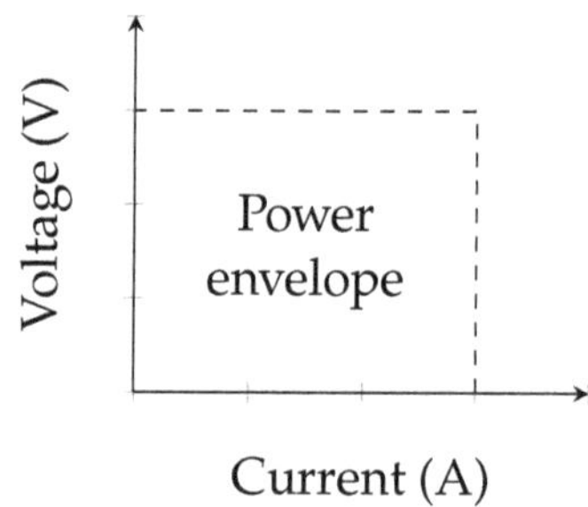

Figure 1.22: Power envelope for a DC supply.

Example 1.9.1: Determine the effect of non-ideal DC supply

If $R_{out} = 10\,\Omega$ in a non-ideal power supply, what will be the output voltage given the DC supply was set to $V_S = 14\,\text{V}$ with $R_{load} = 1\,\text{k}\Omega$ as shown in figure 1.23.

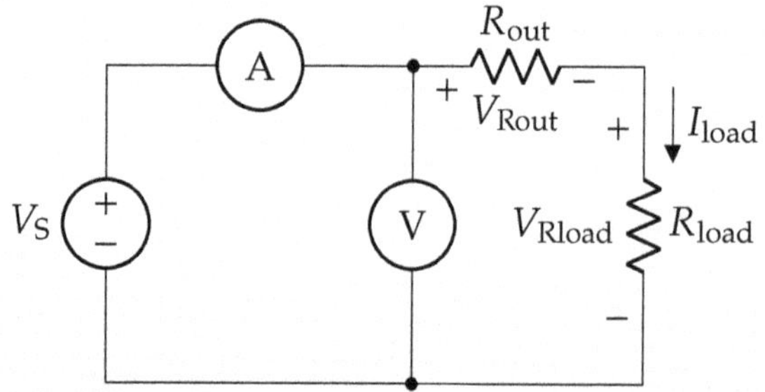

Figure 1.23: DC supply driving a 1 kΩ load.

By applying KVL, we have that:

$$V_S = V_{Rout} + V_{Rload}$$
$$V_S = I_{load} R_{out} + I_{load} R_{load}$$
$$V_S = I_{load}(R_{out} + R_{load})$$

$$I_{load} = \frac{V_S}{R_{out} + R_{load}} = \frac{14}{10 + 1000} = 13.86\,\text{mA}$$

Then, we can find V_{load} by using Ohm's law:

$$V_{load} = I_{load} R_{load} = 13.86\,\text{mA} \times 1\,\text{k}\Omega = 13.86\,\text{V}$$

We expected V_{load} to be 14 V; however, R_{out} only sees 13.86 V. This effect is called loading. The percent error is

$$\frac{|14 - 13.86|}{14} = 1\%$$

Thankfully, most supplies have a low R_{out}.

We can also consider the power lost by the effective output resistance within the supply

$$P_{loss} = I_{load}(V_S - V_{load}) = 13.86\,\text{mA}(14\,\text{V} - 13.86\,\text{V}) = 1.9\,\text{mW}$$

1.10 Tasks

Each experiment in this lab manual contains a set of tasks. Your instructor will determine the specifics of how to record and report your work, but a general guide on some of the key elements of a technical report is included in appendix A on page 211.

Task 1.10.1: DC voltage measurement using a DMM

1. *Adjust* the DC supply to output a 13 V and a −13 V DC. *Measure* both voltages directly with the DMM. *Record* both readings and *compute* the percent error for both readings.
2. Determine how to measure 26 V DC from the DC supply, and then *record* the DMM readings and *compute* the percent error.
3. *Adjust* the supply to output 4.122 V DC. *Measure* with the DMM and *compute* percent error. *Adjust* the power supply output until the DMM measures 4.122 V.
4. *Calculate* the amount of current that will flow through a 330 Ω resistor when 7 V is applied across its leads.
5. *Adjust* the DC supply to 7 V and set the current limit to 15 mA. What do you *expect* will happen if you apply a load of $R_{\text{load}} = 330\,\Omega$? *Record* what happened after applying the load by measuring the voltage across the resistor with the DMM.
6. *Calculate* the minimum resistance that will not exceed the current limit, and then replace the 330 Ω resistor with the calculated value. *Record* the voltage across and current through the resistor using the DMM and the power supply display. *Compare* the precision of each instrument.

Task 1.10.2: Verification of Ohm's Law

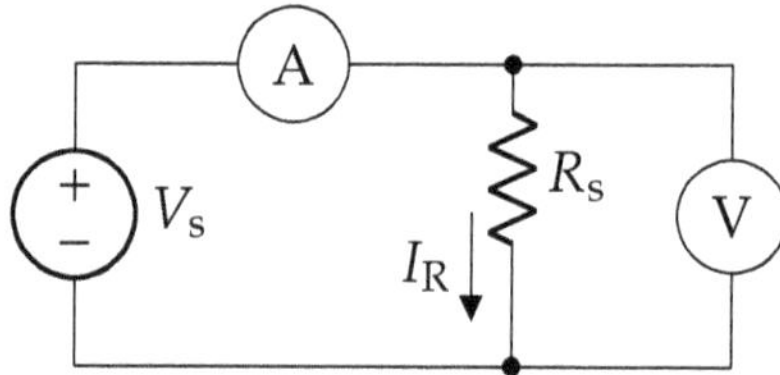

Figure 1.24: Test circuit for verifying Ohm's Law.

1. *Build* the circuit in figure 1.24. Use the DMM as an ammeter and the power supply display as a voltmeter. Pick a value for R_s that is at least 1 kΩ. Set the ammeter to the largest range then decrease the range until an accurate measurement is made.
2. *Set* the current limit on the DC supply to 100 mA.
3. *Adjust* the DC supply output from 0 V to 5 V with a reasonable step size, and *record* I_R from the DMM and V_R from the supply at each step.

4. What do you *expect* the relationship between I_R and V_R to be?
5. *Plot* the I–V relationship of R. What is the best fit relationship between I_R and V_R?
6. *Estimate* the value of R_S using the best fit.
7. *Measure* R_S using the DMM as an ohmmeter.
8. *Calculate* the error between the measured, estimated, and labeled resistances assuming that the DMM is correct.
9. *Calculate* the minimum and maximum resistance the resistor should be based on its tolerance band.

Task 1.10.3: DC current and voltage measurement of a bulb

This task makes use of photoresistors, also known as light dependent resistors (LDRs), to make measurements of light intensity. In short, they change resistance base on the intensity of the light received by them. Typically, the resistance decreases as the light intensity increases. They can be manufactured to be more or less sensitive to certain light wavelengths. The exact relationship for a photoresistor would be reported in the device's datasheet. Datasheets can be found on the manufacturers website as well as in most online parts catalogs.

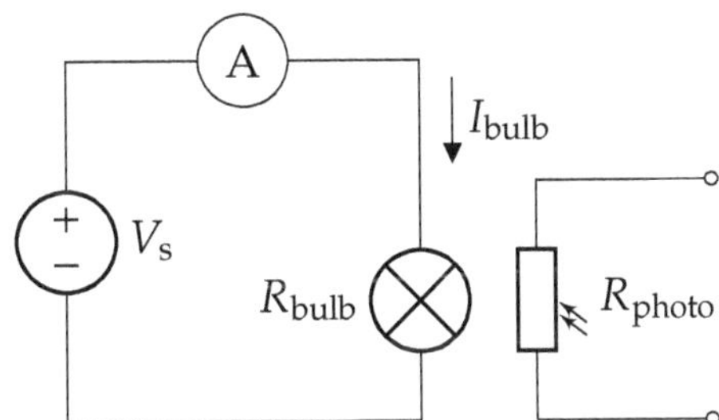

Figure 1.25: Test circuit for measuring different parameters for the light box.

1. *Measure* the provided light bulb's resistance R_{bulb} when it is not yet connected.
2. *Adjust* the DC supply output from 0.1 V to 5 V, and record V_{bulb}, I_{bulb}, and R_{photo} at each step. Cover the light bulb and sensor before taking each resistance measurement.
3. *Compute* P_{bulb} for each measurement.
4. *Plot* I_{bulb} versus V_{bulb}.
5. *Compute* a linear best fit with an intercept at 0 V and 0 A, and plot it over the measured data points. Can the light bulb be accurately modeled as only a resistor?
6. *Plot* R_{photo} versus I_{bulb}. What relationship do you see from the plot?
7. *Plot* R_{photo} versus P_{bulb}. What relationship do you see from the plot?

EXPERIMENT 2

Time-Varying Signals

2.1 Application

Function generators are used to generate *time varying* signals for use within a circuit. Every function generator is different, but most will at least create sinusoidal, sawtooth, triangle, and square waves. Additionally, they typically allow the user to set and adjust, in real time, the signal's frequency, amplitude, phase, and offset. Advanced generators offer significantly more control over the signal shape and frequency content, sometimes they are even capable of generating arbitrary waveforms. Lastly, modern digital function generators can be computer controlled to automate a test procedure. As you can imagine, the function generator is an essential tool for circuit design and verification.

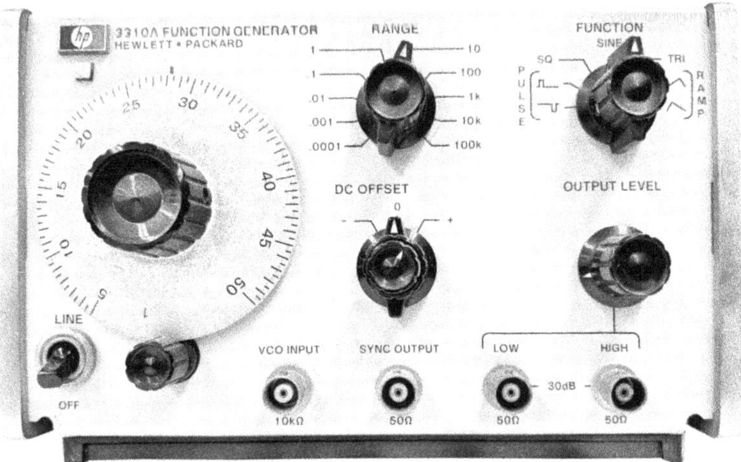

Figure 2.1: A Hewlett-Packard 3310A Function Generator from the early 1970s. It may not have all of the bells and whistles of a modern function generator, but it could still output sine, square, and triangle waves up to 5 MHz.

2.2 AC signals

Properties of an AC signal can change with time, e.g., the voltage may increase and decrease. The set of values that a signal takes at each time is referred to as the waveform. Three frequently encountered periodic waveforms are shown in figure 2.2. Be aware that there are several popular ways of defining the tri and sq functions. Often times one is more convenient to work with for a given problem than the others. In this text, we always use the versions that most resemble a sinusoid.

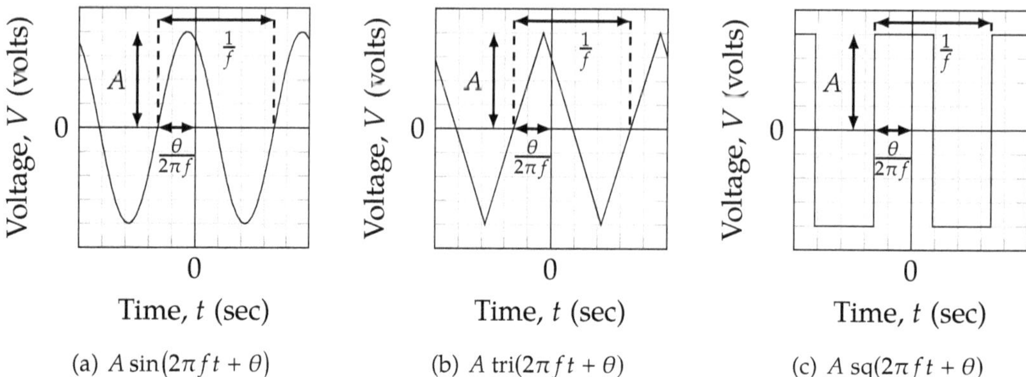

(a) $A \sin(2\pi f t + \theta)$ (b) $A\ \mathrm{tri}(2\pi f t + \theta)$ (c) $A\ \mathrm{sq}(2\pi f t + \theta)$

Figure 2.2: Sine, Triangle, and Square waves

Strictly speaking, AC signals are not necessarily periodic; however, for the purposes of this experiment, we restrict ourselves to only those that are. As a result, the signal can be accurately communicated by specifying the relevant parameters (shape, amplitude, frequency, phase, etc.) of a function. While most modern function generators can generate any arbitrary periodic waveform, the most common waveforms encountered in the lab are the sine, triangle, and square waves. These common waveforms can be specified with only frequency, amplitude, and phase(1).

It is difficult to build intuition about a signal's power with so many parameters, and calculating the power generally involves non-trivial integrals. To simplify the situation, engineers use the root-mean-square (RMS) value.(2)

The primary motivation behind using RMS values in electronics is the fact that a signal with some RMS voltage delivers the same power to a resistive load as a direct, constant in time (DC) signal with a constant voltage equal to the RMS value.(3) For example, a 1 V_{RMS} AC signal heats a resistor the same as a 1 V DC voltage, no matter what shape the AC signal has.

(1) Asymmetric square and triangle waves have one additional parameter called duty cycle and symmetry, respectively. We will look at these parameters in a future experiment.

(2) In fact, it is so common that people often say 120 V AC and assume you understand that it is in volts-rms. Most of the world's power grids are this way. A 120 VAC power feed actually has a peak to peak voltage of about 340 V.

(3) Which, of course, is equal to the DCs RMS value.

2.2.1 Root-mean-square of periodic waveforms

The root-mean-square (RMS) is defined as the square root of the average value of a function squared. In other words, the RMS value of an arbitrary signal $s(t)$ can be computed by

$$s_{\text{RMS}} = \lim_{T\to\infty} \sqrt{\frac{1}{T}\int_{-T}^{T} s^2(t)\,\mathrm{d}t}$$

Given this definition, we can find the RMS value of any signal, including aperiodic signals. However, if we constrain the problem to only periodic signals, we can simplify the limit to equation (2.1).

$$s_{\text{RMS}} = \sqrt{\frac{1}{T}\int_{0}^{T} s^2(t)\,\mathrm{d}t} \tag{2.1}$$

That is, the square root of the average value of the function squared over one period.

Example 2.2.1: RMS value of a sinusoid

The RMS value of a sinusoid can be reduced to a function of just the amplitude. We start by defining the signal, $v(t)$, as

$$v(t) = A\sin(2\pi f t + \theta)$$

where $T = 1/f$. Then, we directly apply equation (2.1)

$$\begin{aligned}
v_{\text{RMS}} &= \sqrt{\frac{1}{T}\int_0^T \left(A\sin(2\pi f t + \theta)\right)^2 \mathrm{d}t} \\
&= \sqrt{\frac{A^2}{2T}\int_0^T \left(1 - \cos(4\pi f t + 2\theta)\right) \mathrm{d}t} \\
&= \sqrt{\frac{A^2}{2T}\left(\int_0^T 1\,\mathrm{d}t - \int_0^T \cos(4\pi f t + 2\theta)\,\mathrm{d}t\right)} \\
&= \sqrt{\frac{A^2}{2T} t \Big|_0^T} \\
&= \frac{|A|}{\sqrt{2}}
\end{aligned}$$

Evaluating equation (2.1) requires using the standard trigonometric identity $\sin^2(t) = \frac{1}{2}\big(1 - \cos(2t)\big)$ and the fact that the integral over a single period of cosine is zero:

$$\int_0^{T=\frac{1}{f}} \cos(2\pi f t + \theta)\, dt = 0$$

It is also possible to compute the RMS value of many different periodic signals in terms of the function's parameters. Some common RMS relationships can be found in table 2.1.

Shape	Function	RMS
DC	$s(t) = A$	$\lvert A\rvert$
Sine wave	$s(t) = A\sin(2\pi f t + \theta)$	$\frac{\lvert A\rvert}{\sqrt{2}}$
Square wave	$s(t) = A\,\mathrm{sq}(2\pi f t + \theta)$	$\lvert A\rvert$
Triangle wave	$s(t) = A\,\mathrm{tri}(2\pi f t + \theta)$	$\frac{\lvert A\rvert}{\sqrt{3}}$

Table 2.1: RMS equations for common periodic signals.[4]

[4] These expressions only apply to AC signals with no DC offset. Equation (2.2) can be used to calculate the RMS value for signals with a DC offset.

2.2.2 Measuring RMS

Most digital multimeters (DMMs) have RMS measurement modes for both voltage and current. Often times this mode is called AC measurement. Multimeters can measure RMS voltages in two ways: DC RMS and AC RMS. True RMS is exactly what you calculate mathematically using equation (2.1). AC RMS measurements first subtract the average value of a signal then calculate the RMS value. In effect, AC RMS is a measure of the energy in the part of the signal that is time varying. Average value, AC RMS and DC RMS are related by equation (2.2), so a DMM will often only measure average value and one of the two RMS values. Always make sure the DMM you are using is measuring the RMS type you expect!

$$s^2_{\mathrm{RMS,DC}} = s^2_{\mathrm{RMS,AC}} + s^2_{\mathrm{AV}} \tag{2.2}$$

2.3 Time-varying Ohm's law

Ohm's law is not a function of time, so when the current and voltage are time varying, we can simply use equation (2.3).

$$v(t) = i(t)R \tag{2.3}$$

If a resistor's voltage is known to be of a certain waveform, then the current must be the same with only the amplitude scaled. The reverse is also true. Using this logic, Ohm's law for an AC signal can be evaluated directly in RMS, eliminating the need to continuously write out the waveform when solving problems!

2.3.1 Time-varying power of a resistor

The DC definition of power can be extended to the time-varying case by allowing both the current and voltage to be time varying, as in equation (2.4).

$$P(t) = i(t)v(t) \tag{2.4}$$

However, similar to the justifications of computing RMS values, it is not intuitive to work with time-varying powers. Instead, the average power is used to describe the energy dissipated in a system. Power in a real resistor is always a positive number because the sign of the current through a resistor is always the same sign as the voltage. Therefore, once can derive a simple equation for average power, starting with equation (2.5).

$$P_{\text{avg}} = \frac{1}{T}\int_0^T P(t)\,\mathrm{d}t \tag{2.5}$$

Using the time-varying version of Ohm's law (equation (2.3)) and power (equation (2.4)), we can write the average power delivered to a resistor as a simple function of the signal's RMS value.

$$\begin{aligned} P_{\text{avg}} &= \frac{1}{T}\int_0^T P(t)\,\mathrm{d}t \\ &= \frac{1}{T}\int_0^T \frac{v^2(t)}{R}\,\mathrm{d}t \\ &= \frac{1}{TR}\int_0^T v^2(t)\,\mathrm{d}t \\ &= \frac{(v_{\text{rms}})^2}{R} \end{aligned}$$

Unsurprisingly, the other versions of the resistive DC power formula hold.(5)

$$P_{\text{avg}} = i_{\text{rms}}v_{\text{rms}} = \frac{(v_{\text{rms}})^2}{R} = (i_{\text{rms}})^2R \tag{2.6}$$

(5) If the load is not purely resistive, these simplifications may not hold. The only way to be sure is to evaluate equation (2.5) directly. For example, inductive and capacitive loads may cause a phase shift between the current and voltage, making $\frac{1}{T}\int_0^T v(t)i(t)\,\mathrm{d}t \neq v_{\text{rms}}i_{\text{rms}}$

2.4 Function generator

When a circuit design requires that an input voltage vary with time, then a function generator is the tool of choice. Function generators are pieces of test equipment that create waveforms over a range of frequencies and amplitudes. Common waveforms include sine, square, triangle, and sawtooth waves. However, they are not all limited to periodic signals. Some are capable of time-limiting the output, while others can output arbitrarily shaped functions fed from stored data or a computer.

By "time-limits" we mean, the signal eventually decays to ground and stays there until the user resets it.

For standard periodic functions, the function generator's interface can be used to set relevant waveform parameters. For example, by selecting a sinusoidal waveform, setting the amplitude to 4 $\mathrm{V_{p\text{-}p}}$[6], offset to 5 V, and frequency to 330 Hz the function generator will output a voltage that follows:

$$s(t) = 2\sin(2\pi 330 t) + 5$$

[6] Be warned, amplitude can be measured in Volts or Volts peak-to-peak. $2\sin(\omega t)$ has an amplitude of 2 V or 4 $\mathrm{V_{p\text{-}p}}$

What about phase? Notice that we did not mention phase. While function generators typically have a phase setting, it is a more advanced feature that will be explored separately. The problem is that phase is a measurement of how a signal is shifted with respect to the "start of time" ($t = 0$ axis); however, there is no clear start of time to use when measuring a single signal. In the context of function generators, we usually use phase to describe how advanced or delayed a signal is *compared* to some other signal. In some circumstances time is synchronized across multiple generators or output channels and phase settings are meaningful.

2.4.1 Equivalent circuit model

Just as with all other devices, function generators have equivalent circuit models that must be considered before connecting them to a circuit. A simple, yet incredibly common, model is shown in figure 2.3. Note that the generator's output current is limited by the internal source resistance. The circuit model is particularly important to understand because function generators typically generate high-quality low-power signals that are easy to load.

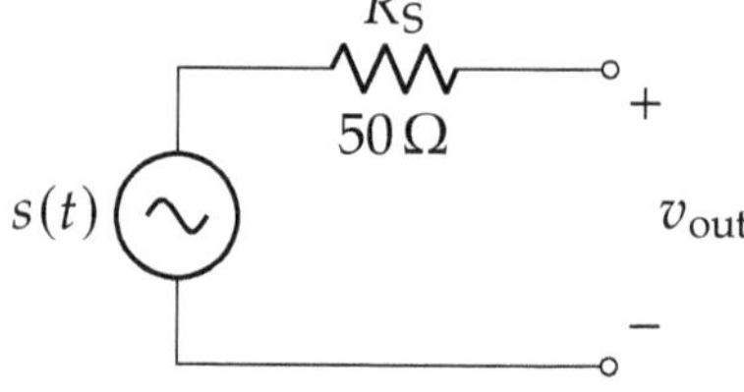

Figure 2.3: Simplified circuit model for function generator.

It may be somewhat presumptuous of us to define the source resistance of *all* function generators as 50 Ω; however, it is nearly an industry-wide standard. The choice of 50 Ω dates back to early microwave radio transmitters of the 1930s as a good a compromise between power handling and low-loss in coax cable.

Example 2.4.1: Function generator loading

Suppose that the function generator in figure 2.4 is configured to generate $5\sin(2\pi 500t)$. What voltage is *actually* delivered to the load?

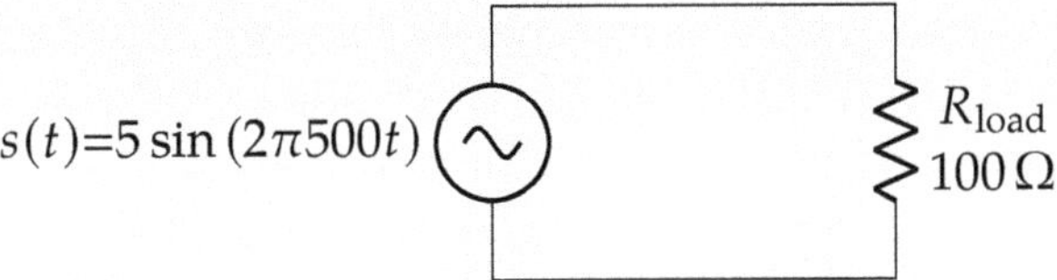

Figure 2.4: Example of "loaded" function generator

Let us start by redrawing the circuit; however, this time we replace the function generator with its circuit model, as shown in figure 2.5.

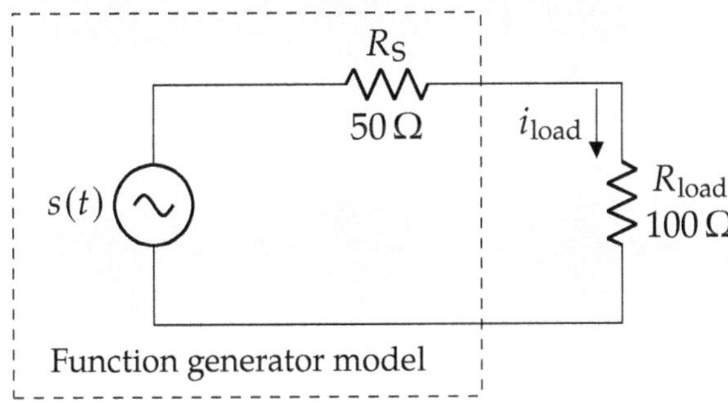

Figure 2.5: Loaded function generator from figure 2.4 replaced with its model.

Now, we use Ohm's law to compute the voltage across the load,

$$
\begin{aligned}
i_{\text{load}} &= \frac{v}{R} \\
&= \frac{s(t)}{R_S + R_{\text{load}}} = \frac{5\sin(2\pi 500t)}{50\,\Omega + 100\,\Omega} \\
&= 33.3\sin(2\pi 500t)\ \text{mA}
\end{aligned}
$$

Therefore,

$$v_{\text{load}} = i_{\text{load}} R_{\text{load}} = 100 i_{\text{load}} = 3.33\sin(2\pi 500t)\ \text{V}$$

Which is about 34% error compared to the expected $5\sin(2\pi 500t)$!

From this result, we see that the function generator's internal source resistance limits the current to something lower than what an ideal generator would develop. This type of reduction of signal between two circuit elements is called *loading* and is a challenge that will often have to designed around. You must always consider the output resistance of the systems you design around to predict what will happen.

2.5 Equipment automation: standard commands for programmable instruments (SCPI)

Modern test equipment can be controlled by computers with, among other technologies, standard commands for programmable instruments (SCPI) [1] [2].(7) SCPI is a standard that defines the syntax, command structure, and data formats to be used by instruments. It is a common industry standard and is well supported by the vast majority of test equipment. Additionally, there are a large amount of available software tools that makes leveraging this interface quite simple.

(7) Pronounced "skippy".

The physical communication bus between the computer and the equipment is not defined by SCPI. In practice, various methods have been used, such as GPIB, RS-232, RS-422, Ethernet, and USB. Thankfully, a widely used software layer called virtual instrument software architecture (VISA) [3] abstracts the details of the physical communication away, allowing the programmer to focus on the control commands. VISA libraries are available in many programming environments including Python, C, MATLAB, and LabVIEW.

For a simple introduction, we will present a very basic description of the SCPI command syntax as well as example MATLAB code sessions. While not an end all reference, it should be enough to get you started.

2.5.1 SCPI command syntax

All SCPI commands are a set operation, a query operation, or both. Set operations tell the instrument to do something or change a setting. Query operations request data from the equipment. A question mark at the end of a query command indicates that a response is wanted (otherwise, the equipment may not send anything back in return). Your equipment's manual should list all of the supported commands. For the remainder of the example, we will use a Keysight 34405A DMM; however, the commands are likely to work on any DMM that supports SCPI with no or minor modification.

Common commands

There are some SCPI commands that the standard requires all equipment to implement. Common commands start with an asterisk. For now, we will only need one of them: the identification command.

The identification command is `*IDN?` and when sent will cause the equipment to respond with a string similar to

```
Keysight Technologies,34405A,<Serial Number>,II.II-MM.MM
```

This is handy when automatically identifying the series of instrument you are speaking with.

Measurement commands

Commands are grouped into a collections of sub-commands and then stacked into a hierarchy, or tree. For example, the `:MEASure` group contains all of the measurement sub-groups. Sub-groups are named by concatenating the path in the tree together with colons (:). Below is an example of a relatively simple `:MEASure` group

```
:MEASure
    [:VOLTage]
        :AC? [{<range>|AUTO|MIN|MAX|DEF} [,{<resolution>|MIN|MAX|DEF}]]
        [:DC]? [{<range>|AUTO|MIN|MAX|DEF} [,{<resolution>|MIN|MAX|DEF}]]
    :CURRent
        :AC? [{<range>|AUTO|MIN|MAX|DEF} [,{<resolution>|MIN|MAX|DEF}]]
        [:DC]? [{<range>|AUTO|MIN|MAX|DEF} [,{<resolution>|MIN|MAX|DEF}]]
    :RESistance? [{<range>|AUTO|MIN|MAX|DEF} [,{<resolution>|MIN|MAX|DEF} ]]
    :CONTinuity?
```

Category	Symbol	Meaning
Braces	{}	Encloses the parameter choices for a given command string
Vertical Bar	|	Separates multiple parameters
Triangle Brackets	<>	Indicates that you must specify a value for the enclosed parameter
Square Brackets	[ ]	Indicates that the parameter is optional and can be omitted if not needed

Table 2.2: SCPI Syntax

The command syntax is to be interpreted by using the conditions in table 2.2. It may look intimidating, but the examples should help break it down. The CAMel case indicates the short and long version of the command, e.g., `MEAS` and `MEASURE` are the same command and either is valid. The case of the commands does not matter, however, it is standard practice to capitalize everything.

As an example, the full command to make a DC voltage measurement with AUTO range and maximum resolution is

```
MEASURE:VOLTAGE:DC? AUTO, MAX
```

However, using the rules from table 2.2, the command could be shortened to

```
MEAS?
```

This is due to using the short form command, dropping the optional `:VOLTage` and `:DC` sub-commands, and relying on the default values for the parameters (AUTO and MAX, respectively).(8)

(8) You should consult your equipment's programming reference or manual to verify command definition and defaults.

As a final example, the shortened command for a resistance measurement using AUTO ranging and MAX resolution is

```
MEAS:RES?
```

2.5.2 Example MATLAB session

Connecting to a piece of equipment in MATLAB is simple.(9)

(9) Other languages such as Python have VISA bindings that are at least as easy as the following. We encourage you to explore those free and open source options.

We start by asking MATLAB to issue a reset to the instrument system. While technically not required, it is a good habit.

```
>> instrreset;
```

To connect to the instrument, we need to know its address. This can be found using a tool like Keysight Connection Expert or in MATLAB using `instrhwinfo`.

```
>> visainfo = instrhwinfo('visa','keysight');
>> visainfo.ObjectConstructorName
```

This will print the commands needed to connect to all of the devices currently detected. Once we know the address of the device, we create a visa handle using the `visa` command followed by the `fopen` command to connect.

```
>> instr = visa('keysight', 'USB::0x1234::567::A89-0::INSTR');
>> fopen(instr);
```

At this point, we can send and receive messages from the device. To verify that we connected to the correct device, we send the `*IDN?` VISA command using `fprintf`. The response is read using `fscanf`.

```
>> fprintf(instr, '*IDN?');
>> response = fscanf(instr);
>> fprintf("Connected to %s\n", response);
```

Finally, we can take measurements with the equipment. We'll need to convert the string output into a number using `str2double`. We could also use a format string with `fscanf`.

```
>> fprintf(instr, 'MEASURE:VOLTAGE:DC? AUTO MAX');
>> voltage = str2double(fscanf(instr));
>> fprintf("Measured voltage: %.4f", voltage)
```

Lastly, to properly cleanup the resources and restore the state of the equipment, we need to disconnect. To do that, we use `fclose` to end the connection and `delete` to remove the visa handle from memory.

```
>> fclose(instr)
>> delete(instr)
```

For reference, the same session can be accomplished in Python using PyVISA[4]. The following is a copy of a Python interpreter session:

```
>>> import visa
>>> rm = visa.ResourceManager()
>>> rm.list_resources()
('USB::0x1234::567::A89-0::INSTR')
>>> instr = rm.open_resource('USB::0x1234::567::A89-0::INSTR')
>>> voltage = float(instr.query('MEASURE:VOLTAGE:DC? AUTO MAX'))
>>> print("Measured voltage: {.4f}".format(voltage))
>>> instr.close()
```

2.6 Prelab

Task 2.6.1: Prelab questions

1. *Make* a computer generated plot of a triangle wave with an amplitude of 3 V, offset of 1 V, and frequency of 1 kHz; that is:

$$s(t) = 3\,\text{tri}\left(2\pi 1000t\right) + 1 \tag{2.7}$$

2. *Compute* the DC RMS and AC RMS voltage value of equation (2.7).
3. *Locate* the SCPI reference for your function generator. It is often located in either the manual, datasheet, or programming reference on the manufacturer's website. Record the document's URL or location information.
   a) Read through the source for the `fg_sin` and `fg_output` functions provided.[a]
   b) In your own words, explain what each function does.
   c) Using the manual or programming reference for the function generator, explain what the following SCPI commands do:
      - `SOUR1:FUNC SIN`
      - `SOUR2:FREQ 1000`
      - `SOUR1:VOLT 1.5`
      - `SOUR2:VOLT:OFFS 3`
      - `OUTP2 ON`

[a] Electronic resources can be found on your course website.

2.7 Tasks

Task 2.7.1: Root mean square value measurement

1. Use the function generator to *generate* a sine wave with an amplitude of 2 V at 1 kHz with a DC offset of 4 V.
2. *Compute* the AC and DC root mean square value of the sine wave.
3. *Measure* the RMS voltage and the frequency of the sine wave using the DMM.
4. *Compare* the computed root mean square value with the measured one. Do they agree with each other? Does the DMM measure AC RMS or DC RMS? Does the measured frequency agree with the theoretical value? *Compute* the percent error.
5. *Repeat* the same procedures using a triangle wave with an amplitude of 3 V, a frequency of 2 kHz, and an offset of −2 V.

Task 2.7.2: Automation library: RMS measurement

Throughout these experiments you will be asked to develop various measurement automation techniques. To simplify the more complicated tasks, you should develop a library of useful functions. Please keep these functions available for future experiments. To complete this task, you will need to use the procedure given in section 2.5.2 to connect to your DMM. Make sure it is on before attempting to connect.

1. *Write* a MATLAB or Python function, `dmm_rms_volt`, that accepts a VISA handle. It should make all required measurements and then return the DC and AC RMS voltages.

   MATLAB:

   ```
   function [dc_rms, ac_rms] = dmm_rms_volt(instr)
       ...
   end
   ```

   Python:

   ```
   def dmm_rms_volt(instr):
       ...
   return (dc_rms, ac_rms);
   ```

   Consult the SCPI reference manual of your DMM for the command to make a resistance measurement.

   Note: By default the output of `fscanf` is a string. You should use the `formatSpec` parameter to cast the string into a double. Run `help fscanf` for more info.

2. *Use* your function to find the RMS values of the signals from task 2.7.1. Does the function work as expected? Include your completed functions in your report.

Task 2.7.3: Automation library: Function generator output

To simplify the more complicated tasks, a library of SCPI commands will be provided to help automate experiments. You can utilize the provided functions to control the equipment for automation tasks. Review the source code of each function before using it in order to get an idea of how these functions work.

1. *Read* the MATLAB or Python functions, `fg_sin` and `fg_output`. [a]

   `fg_sin` is a function that accepts a VISA handle, channel number, frequency, volts peak-to-peak, and offset. It configures the function generator's output shape (sine), frequency, volts peak-to-peak, and offset on the given channel. The function does not turn the channel on or off.

   ```
   fg_sin(instr, channel, freq, vpp, offset)
   ```

`fg_output` is a function that accepts a VISA handle, channel number, and state. The state value can either be `ON` or `OFF`. The function sets the state of the given channel's output.

```
fg_output(instr, channel, state)
```

2. *Use* the automation library to set channel 1 to $0.75\sin(2\pi 1400t) + 1$ and channel 2 to $1.4\sin(2\pi 60t)$. Does the function generator display indicate the correct outputs?
3. *Use* the automation library to turn both channel 1 and 2 on. *Measure* the RMS voltage of each channel. Does the measurement agree with your expectations?
4. Copy your MATLAB or Python session or script into your report.

[a] Electronic resources can be found on your course website.

Task 2.7.4: Verification of Ohm's Law using an AC voltage source

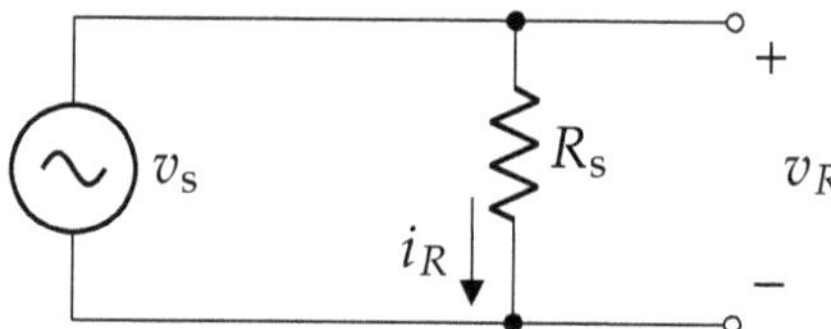

Figure 2.6: Test circuit for verifying Ohm's Law using a AC voltage source.

1. *Build* the circuit in figure 2.6. Pick a value of $R_S > 1\,\text{k}\Omega$.
2. *Configure* the function generator to generate a sine wave with the following properties:
   - Peak to peak voltage of 10 V
   - Offset of 0 V
   - Frequency of 100 Hz
3. *Apply* the signal to the circuit and then *measure* the RMS values of i_R and v_R using the DMM.
4. *Plot* i_R versus v_R by varying the amplitude of the sine wave between 0 V and 5 V. Does Ohm's Law still hold with an AC voltage source?

Task 2.7.5: Series and Parallel Bulbs

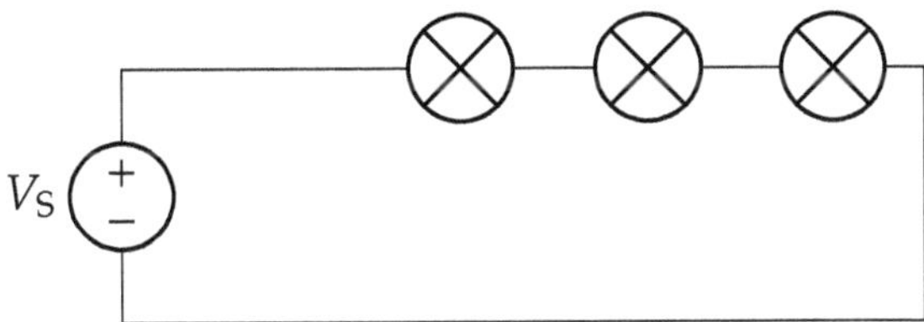

Figure 2.7: Test circuit when all light bulbs are in series.

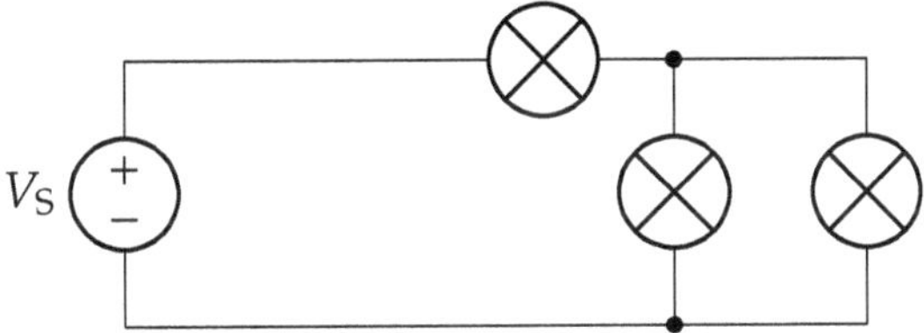

Figure 2.8: Test circuit when two light bulbs are in parallel and one in series.

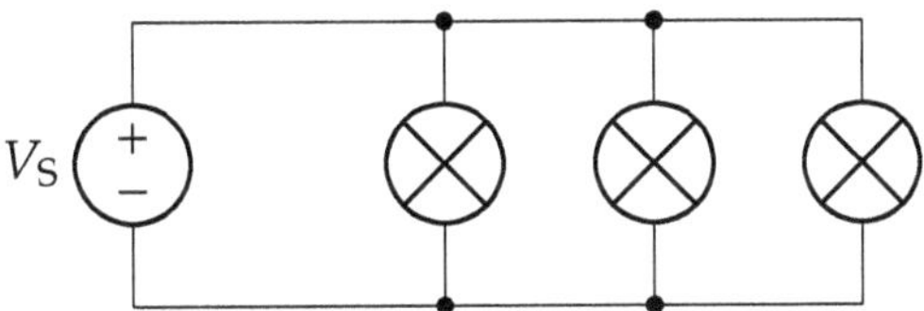

Figure 2.9: Test circuit when all light bulbs are in parallel.

1. For each test circuit from figure 2.7 to figure 2.9:
   a) *Build* the circuit. Use the power supply as the DC source.
   b) *Apply* a 5 V DC voltage to the circuit.
   c) *Measure* each bulb's voltage and current with the DMM.
   d) *Measure* the resistance, R_p, of a photo resistor placed directly next to and facing each bulb.
   e) *Compute* each light bulb's P_{avg}.
2. *Explain* the difference in brightness of the bulbs for each circuit configuration. Why does the brightness change with the wiring?
3. One of the configurations is the same as the configuration for power distribution in a modern building. Which one do you think it is? Why?
4. Replace the DC power supply with the function generator for the circuit in figure 2.9. Configure the function generator to output DC 5 V. *Record* the voltage across the bulbs and the current through a bulb.
5. Compare the power delivered to the bulbs between the function generator and power supply set to the same voltage. Why are the values different?

2.8 References

[1] *SCPI consortium.* [Online]. Available: http://www.ivifoundation.org/scpi/default.aspx.

[2] *Standard commands for programmable instruments volume 1: Syntax and style,* May 1999. [Online]. Available: http://www.ivifoundation.org/docs/scpi-99.pdf.

[3] *IVI specifications.* [Online]. Available: http://www.ivifoundation.org/specifications/default.aspx.

[4] *PyVISA.* [Online]. Available: https://pyvisa.readthedocs.io/en/latest/index.html.

EXPERIMENT 3

The Oscilloscope

3.1 Application

The oscilloscope is the primary tool for viewing time-varying waveforms in a laboratory environment. They can be used to view waveforms that change far faster than the human eye can see. Oscilloscopes are also useful for viewing the minute details of signal that could not be determined from a digital multimeter (DMM).

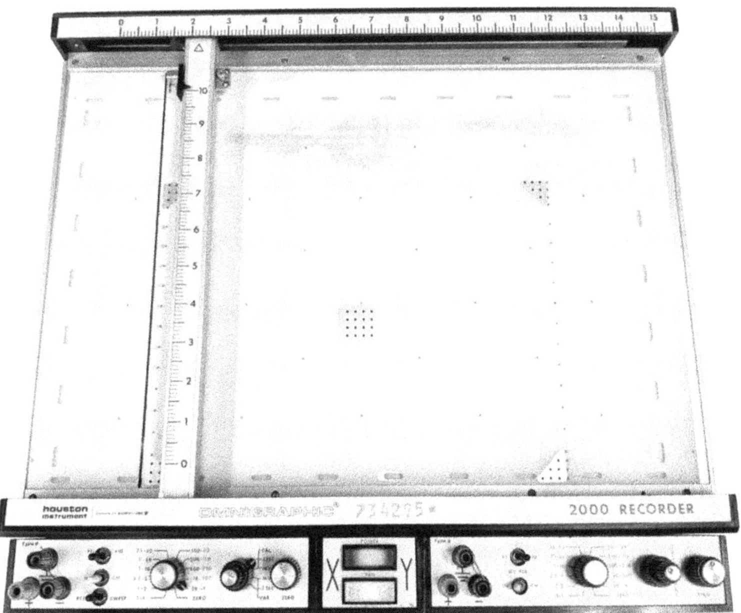

Figure 3.1: A voltage controlled plotter that could be used to record waveforms in real time. This device could be used to record a measured waveform before it was easy to save and print oscilloscope displays.

3.2 Horizontal and vertical controls

The horizontal and vertical controls on the oscilloscope are used to obtain the optimum view of the waveforms being measured. Most oscilloscopes have a single horizontal setting that applies to all signals and different vertical controls for each channel.

Vertical offset The vertical offset moves the displayed signal up and down on the screen. It allows for vertically centering signals with a DC offset and moving signals around to conveniently display multiple signals without overlap. Vertical offset can also be used to align signals with the division markers on the oscilloscope screen.

Time offset The time offset moves the displayed signal left and right on the screen. It is useful for inspecting different portions of a signal with a period longer than what can be viewed conveniently on the screen. Time offset can be used to align the signal with the division markers on the screen for easier measurement.

Vertical scale Vertical scale stretches the signal on the screen. To capture the most detail, use the largest vertical scale in which the entire signal can be seen. When the vertical scale is too large or too small, the oscilloscope may not display the signal properly, and automated measurements may not be correct.

Time scale The time scale sets how much time is displayed on the screen. Adjusting it will stretch or compress the signal horizontally. Time scale can be used to set how many periods of a signal are visible at once.

Autoscale All modern oscilloscopes have some form of autoscale or autoset. When this button is pushed, the oscilloscope will attempt to find the best settings for displaying the waveforms at the inputs. Autoscale is not always correct. When it fails to align the waveforms as needed, the horizontal and vertical controls must be used to fine tune the display. Autoscale will also reset any horizontal or vertical settings. Many simple signals can be easily displayed using the autoscale feature, but it is important to be able to correctly set horizontal and vertical settings in lieu of autoscale because many more complicated signals will not be displayed correctly using only autoscale.

Autoscale is a very useful tool, but all good engineers should be able to acquire and view a mystery signal without using autoscale.

3.3 Signal viewing modes

Oscilloscopes have several viewing modes that can be useful in different applications. The modes used in this lab manual are roll mode, normal mode, and XY mode.

Normal Mode Normal mode is the default view. For each trigger event(1) that occurs, the oscilloscope will plot enough data to fill the screen. Autoscale will always put the display into normal mode, so if a different mode is needed, then autoscale cannot be used.

(1) Trigger events tell the oscilloscope to record data. We will go into detail on triggering in the next experiment.

Roll Mode Roll mode plots all measured data points as soon as they are measured. It is useful for measuring slow signals or viewing how a signal changes over longer periods of time. It is called roll mode because the signal "rolls" across the oscilloscope screen.

XY Mode XY mode plots one input on the X axis and the second input on the Y axis. It is most useful when the relationship between two voltages is needed. Even though there is no time axis in XY mode, the time scale control will vary the update rate of XY mode. Some oscilloscopes also have an XYZ mode, where the third input adjusts the brightness of the trace(2).

(2) XYZ mode has fun applications like turning an oscilloscope into a TV, like shown in this video: https://youtu.be/5FYF5uhCzAM

3.4 Measurement tools on the oscilloscope

The oscilloscope provides a variety of tools used to measure different aspects of a signal. The advanced features may vary greatly between different oscilloscopes, but all modern oscilloscopes will have the following basic features.

Cursors The cursor is the most basic measurement tool on the oscilloscope. The X cursors are used to measure time in normal and roll mode and voltage in XY mode. The Y cursors are used to measure voltages in all modes. When in normal or roll mode, the Y cursors will only be correct for a single input unless the offset and scale are the same for all signals. The oscilloscope has settings to choose which input the Y cursor applies to.

Automatic measurement Oscilloscopes can measure different aspects of a signal automatically without using cursors. Common measurements include peak to peak voltage and frequency, but there are many more options in the measure menu. Take time to look through the various auto measure

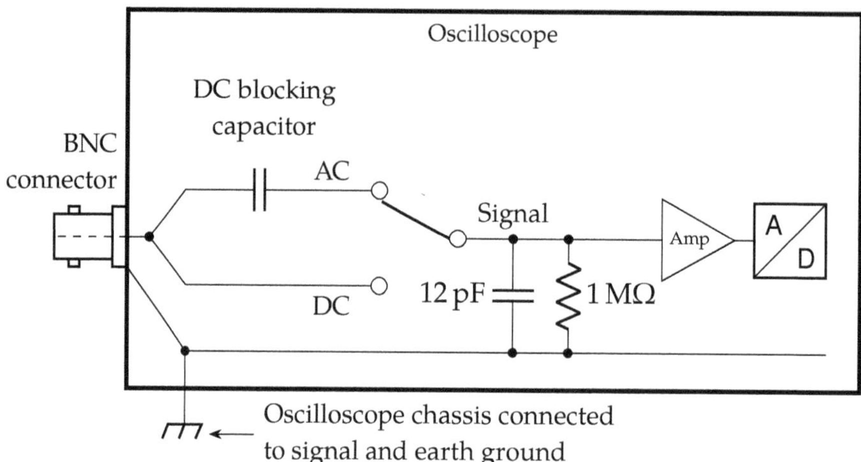

Figure 3.2: Example input stage of an oscilloscope

commands present on your oscilloscope. They may be useful when debugging complicated circuits.

Automatic measurements may not be correct if the signal is not fully displayed on screen with the largest acceptable vertical scaling. Horizontal scaling should be set so that the screen shows an appropriate amount of the signal with respect to the measurement type.

Math It is ocassionally useful to be able to perform some simple math operations on a signal or set of signals. Most oscilloscopes have the ability to add, subtract, and multiply signals on different inputs. Other more advanced math features are typically available, but the options may vary between different oscilloscopes.

3.5 The internals of an oscilloscope

The input stage of an oscilloscope is shown in figure 3.2. There are several parameters that are important to consider when measuring circuits with the oscilloscope. The input of the oscilloscope is measured across a resistor and capacitor. This value is typically printed on the front of the oscilloscope near the inputs. A typical oscilloscope will have an input capacitance and resistance on the order of 12 pF and 1 MΩ, respectively.

The input resistance is important if the Thévenin equivalent resistance of the circuit is close to or above 1 MΩ. When that happens, a voltage divider is formed and the oscilloscope will measure a lower voltage than was expected. The input capacitance will limit the frequency of signals that can be measured using the oscilloscope. At very high frequencies, a capacitor acts like a short circuit, so it will cause the oscilloscope to measure no voltage at the input.

The input resistance and capacitance can be adjusted by using an x10 probe. An x10 probe will multiply the resistance by 10 and divide the capacitance by 10. This allows for measurement of higher frequency signals or circuits with a high Thévenin equivalent resistance.

The AC coupling setting of the oscilloscope can be used to remove the DC offset of a signal. If the DC offset of a signal needs to be measured, then DC coupling must be used.

The names DC and AC coupling may appear to be a bit of a misnomer as a time varying (AC) signal can be measured using DC coupling, but a constant signal cannot be measured using AC coupling.

3.6 The voltage divider

A voltage divider reduces the voltage amplitude of a signal and can be understood with Ohm's law. For the two resistor divider shown in figure 3.3, we have a total equivalent resistance of $R_{eq} = R_1 + R_2$ because R_1 and R_2 are in series. Using Ohm's law, the total current in the circuit is equal to

$$I = \frac{v_{IN}}{R_{eq}} = \frac{v_{IN}}{R_1 + R_2}$$

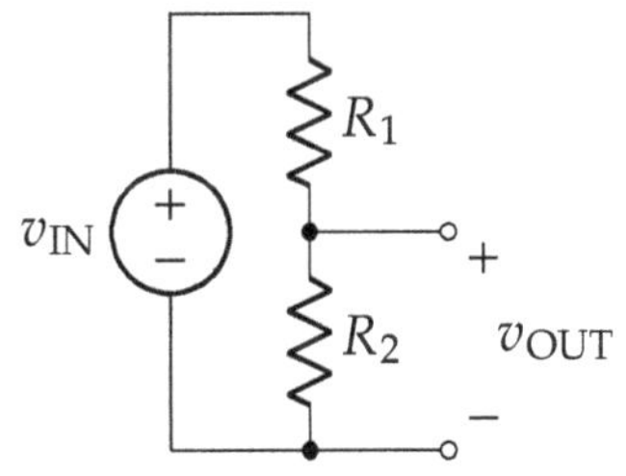

Figure 3.3: Voltage Divider Circuit

Then, v_{OUT} is computed with Ohm's law on R_2

$$v_{OUT} = I \times R_2 = v_{IN} \times \frac{R_2}{R_1 + R_2} \tag{3.1}$$

Equation (3.1) can be solved in reverse to find values of R_1 and R_2 for a certain input to output voltage ratio[(3)].

[(3)] This ratio is called attenuation when the output voltage is less than the input.

The potentiometer A common circuit component that implements a voltage divider is the potentiometer. A potentiometer has three terminals. The outer two connect to two ends of a resistor and the middle terminal connects to a wiper that slides across it. The potentiometer circuit in figure 3.4 is a common manually adjustable voltage attenuator.[(4)]

[(4)] This circuit generates voltage for signal purposes, but be careful—the Thévenin impedance of this source may be high and should not be used as a power supply. The correct device for generating a lower supply voltage than the input voltage is called a voltage regulator, and we will explore them in future experiments.

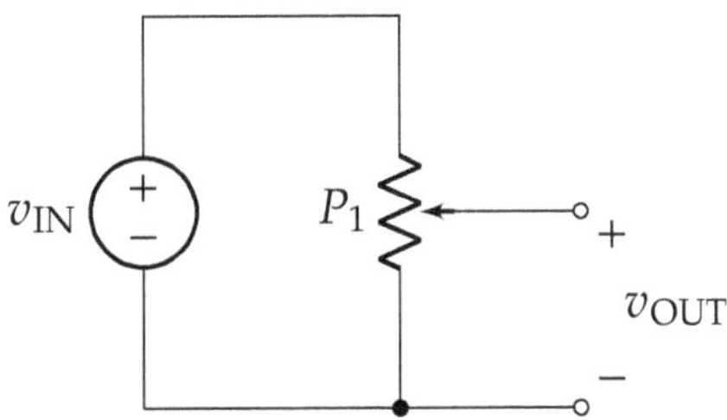

Figure 3.4: A potentiometer based voltage divider circuit

3.7 Prelab

Task 3.7.1: Prelab Problems

1. *Watch* the introduction video for this experiment. What are three new things you learned about oscilloscopes?
2. A sinusoidal signal has a peak to peak voltage of 7 V, an offset of 2 V, and a frequency of 600 Hz. *Write* this signal in the form $s(t) = A\cos(2\pi f t) + B$.
3. Which oscilloscope mode should be used for very slow signals?
4. Using the circuit in figure 3.3, *calculate* v_{OUT} if $R_1 = 25\,\text{k}\Omega$, $R_2 = 75\,\text{k}\Omega$, and $v_{\text{IN}} = 10\,\text{V}$. Next, *calculate* v_{OUT} when a $1\,\text{k}\Omega$ resistor is connected across the output.[a]
5. *Do* the Analog Discovery Introduction on the course website.

[a] Hint: the 1 kΩ and 75 kΩ resistors are in parallel.

3.8 Tasks

Task 3.8.1: Analog Discovery Calibration

1. *Calibrate* your Analog Discovery 2 by following the tutorial on the course website.

Task 3.8.2: Horizontal and vertical scale 

1. *Set* the function generator to generate a sinusoid with the following properties:
   - Peak to peak voltage of 2 V
   - Offset of 0 V
   - Frequency of 1 kHz
2. *Write* the signal generated in the form $v(t) = A\cos(2\pi f t)$.
3. *Connect* the output of the function generator to the input of the oscilloscope using a probe set to "1x" mode, then press the oscilloscope's autoscale[a] button.
4. *Adjust* the horizontal scale so that four periods of the sinusoid fits on the display.
5. *Adjust* the vertical scale to 1 V per division.
6. *Measure* the peak to peak voltage and frequency using the oscilloscope's cursors. *Calculate* the error for these measurements.
7. *Capture* a screenshot of the oscilloscope display including the cursors.

[a] This may also be called "autoset."

Task 3.8.3: Horizontal and vertical position 

1. *Set* the function generator to generate a triangle wave with the following properties:
    - Peak to peak voltage of 2 V
    - Offset of 4 V
    - Frequency of 5 kHz
2. *Connect* the output of the function generator to the input of the oscilloscope.
3. *Adjust* the vertical position and scale so that the ground icon is visible on the display and the minima and maxima of the triangle wave are aligned with the division markers.
4. *Adjust* the horizontal position so that the leftmost part of the signal is at a peak of the triangle wave.
5. *Measure* the peak to peak voltage, frequency, and average value using the oscilloscope's measurement tools. *Calculate* the error for these measurements.
6. *Capture* a screenshot of the oscilloscope display that includes the measurements.

Task 3.8.4: Slow Signals and Roll Mode

1. *Set* the function generator to generate a sawtooth wave with the following properties:
    - Peak to peak voltage of 4 V
    - Offset of 2 V
    - Frequency of 0.2 Hz
    - Increasing with time
2. *Connect* the output of the function generator to the DMM. Set the DMM to DC measurement mode. Are you able to plot the signal by hand using the DMM readings?
3. *Connect* the function generator back to the oscilloscope. Try to use autoscale to view the signal. Does autoscale get the settings correct?
4. *Set* the oscilloscope to roll mode[a].
5. *Adjust* the horizontal and vertical controls until an entire period of the waveform is visible, the waveform is centered vertically, and the waveform is four divisions tall.
6. *Press* the Run/Stop or Single button, then *capture* a printout of the oscilloscope display. Circle the horizontal and vertical scale used.

[a] This is typically found in the horizontal controls

Task 3.8.5: Automation Library: Oscilloscope display (AD2)

Two MATLAB or Python functions will be provided for this experiment:
`scope_time_setup(instr, scale, position)` sets the time axis seconds per division to `scale` and the zero time offset to `position`.
`scope_input_setup(instr, channel, scale, offset)` sets the vertical axis volts per division to `scale` and the vertical voltage offset to `offset` for the input `channel`.

1. *Write* a function

   ```
   display_sin(fg, scope, channel, frequency, amplitude, offset)
   ```

   that does the following:

   - Sets the function generator to output a sinewave with amplitude `amplitude`, offset of `offset`, and frequency of `frequency` on output `channel` (use `fg_sin` from experiment 2).
   - Sets the time scale of the oscilloscope to display 2 periods of the sinewave.
   - Sets the vertical scale and offset of channel `channel` on the oscilloscope so that the sinewave is at least four divisions peak to peak and vertically centered.

2. *Test* the function for three different pairs of frequency, amplitude, and offset. Include a copy of the functions in your report. You do not need a procedure or conclusion for this section of the report.

Task 3.8.6: Voltage Division

1. *Set* the function generator to generate a sine wave with the following properties:
   - Peak to peak voltage of 4 V
   - Offset of 0 V
   - Frequency of 2 kHz
2. *Build* the voltage divider circuit shown in figure 3.4 using a 10 kΩ potentiometer. *Connect* the input of the circuit to the function generator and the first input on the oscilloscope. *Connect* the output of the circuit to the second input on the oscilloscope.
3. *Adjust* the knob on the potentiometer. Describe what happens to the signal.
4. *Set* the potentiometer so that the output signal displayed on the oscilloscope is 1 V peak to peak. *Capture* and oscilloscope screenshot of the waveform that clearly shows the input and output with the same vertical scale.

EXPERIMENT 4

Oscilloscope Triggering and Coupling

4.1 Application

Not all signals can be correctly viewed by simply pushing autoscale. These signals require extra settings in the oscilloscope trigger menu. The triggering and coupling settings on an oscilloscope expand the versatility of it as a measurement instrument when dealing with complex waveforms.

Fully understanding the trigger settings of an oscilloscope enables you to set up complicated test environments that can catch errors and anomalies in circuit performance without hoping that you can catch a glitch at just the right time. The effective use of the triggering system in an oscilloscope will make you more efficient during the design and debug phases of the circuits you design.

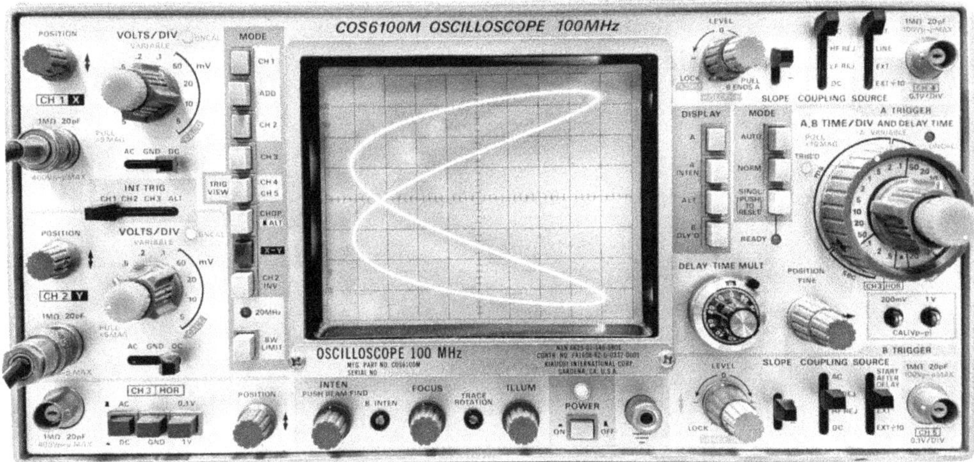

Figure 4.1: A COS6100M Oscilloscope displaying a Lissajous figure in XY mode.

4.2 Input coupling

The input stage of an oscilloscope commonly has at least two settings: AC and DC coupling.

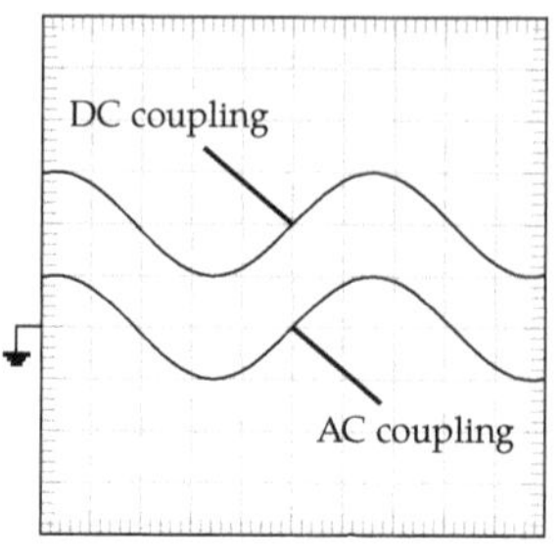

Figure 4.2: The signal $2 + \sin(t)$ with AC and DC coupling

DC coupling DC coupling passes the unabridged signal to the oscilloscope input. This enables viewing and measuring the entire signal, assuming it is within the oscilloscope's allowable voltage range.

AC coupling AC coupling passes the signal through a capacitor to remove any DC offset voltage before going to the oscilloscope input. Even though AC coupling removes information from the signal, it is useful when inputs have large or changing offsets. It is not uncommon to be interested in only the variation around a certain DC signal, such as measuring noise on a power supply or individual bits of a digital signal.

Many measurements of mid– to high– frequency signals are more conveniently done with AC input coupling because the DC level is often ignored in performance metrics for such systems. On the other hand, if you are interested in slow changes in offset voltage, or even in the offset voltage itself, then DC coupling is necessary.

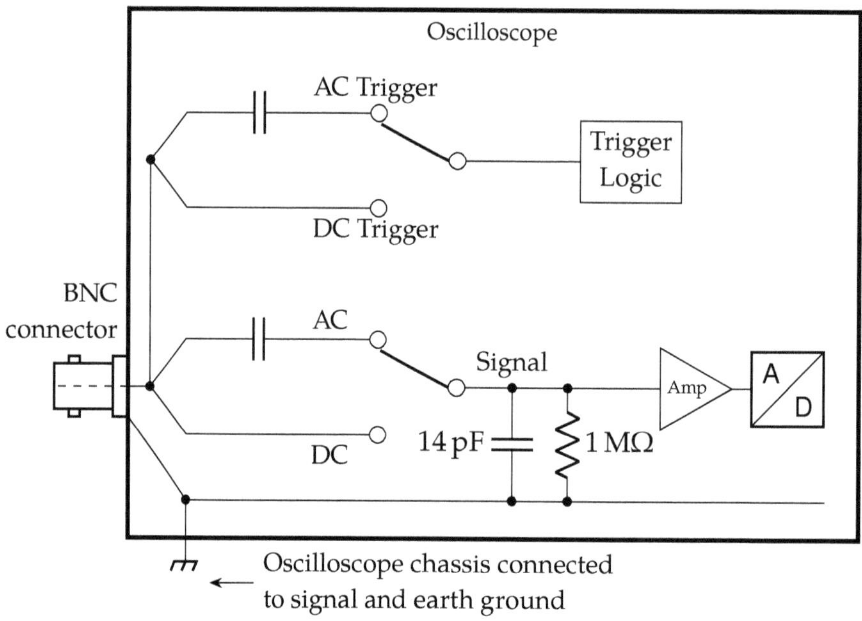

Figure 4.4: Example input stage of an oscilloscope with trigger coupling shown

4.3 Triggering

The triggering settings on the oscilloscope determine when the oscilloscope starts recording data. The most common triggering method used is edge triggering. Whenever the signal crosses the set trigger level, the oscilloscope will record a small duration in time and display it on the screen. Oscilloscopes can be set to trigger on a rising or falling edge. Rising edge triggers the recording when the signal goes from less than the trigger level to above it, and falling edge does the opposite.

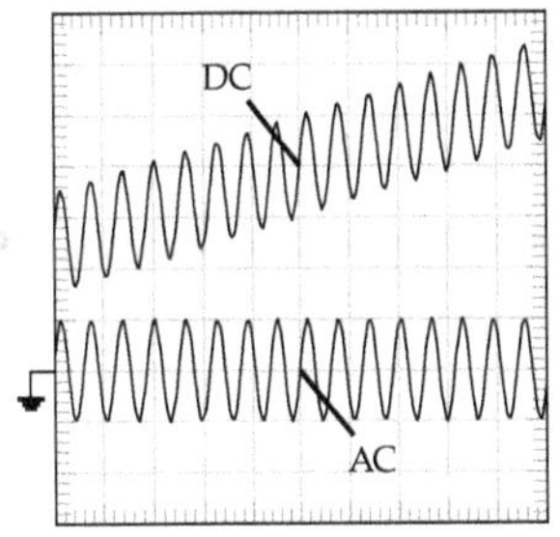

Figure 4.3: The signal $\sin(10t)$ with a slowly changing DC offset shown with AC and DC coupling

There are a wide variety of triggering options available on most modern oscilloscopes. You are encouraged to explore the variety of modes and options on your particular oscilloscope and on the Analog Discovery 2.

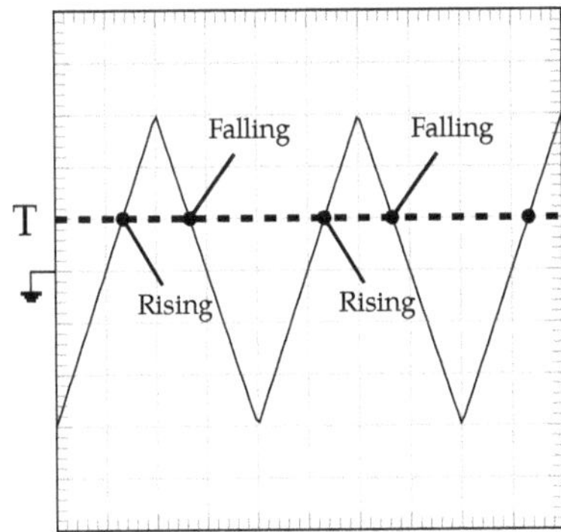

Figure 4.5: Edge trigger locations of a $6\,V_{p\text{-}p}$ triangle wave with trigger level at 1 V

There are two common data acquisition modes for triggering: auto and normal.

Auto mode Auto mode will occasionally trigger even if the trigger condition set is not met.[1] This mode is useful for looking at the shape of a signal to determine the proper trigger settings.

[1] Oscilloscopes vary, but often the auto trigger occurs after a certain length of time passes without a trigger caused by the input.

Normal mode Normal mode only triggers when the set condition is met. When the condition is not met, the display will either be blank or frozen on what was shown from the last trigger. It is important to locate the symbol on the scope that indicates triggering is occurring. It can be quite confusing when the oscilloscope shows a signal even after disconnecting the input!

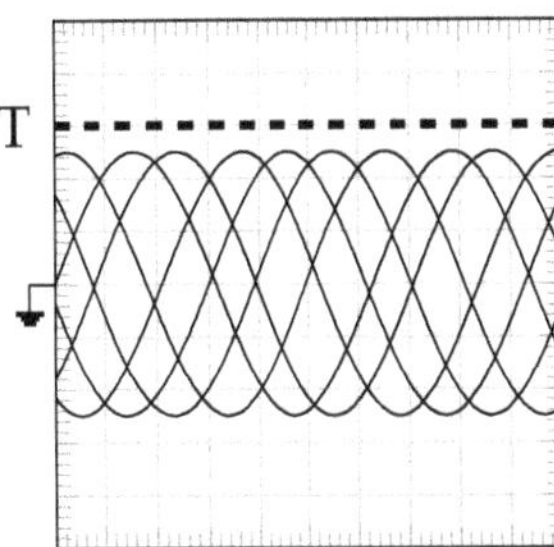

Figure 4.6: Scope display in automatic trigger mode when the signal is not correctly activating the trigger condition

In auto mode, an incorrectly triggered signal will appear like an overlay of many phase shifted copies of the actual signal. This is due to the scope falling back to artificial triggering, resulting in copies being draw on the screen from wherever the signal is at that moment. It is likely that the trigger rate is not an integer multiple of the signal's period, resulting in most of the drawings being out of phase with the rest. See figure 4.6 as an example.

4.3.1 Triggering with multiple inputs

Normally an oscilloscope will only trigger on one input, even if more than one is in use. The trigger event causes the scope to sample and draw all of the signals at once. Therefore, unless the other inputs have periods with exact multiples of the signal being triggered on, they may appear unstable. Similar to auto mode with automatic trigger events. Function generators have special outputs (sometimes called "sync") that can help remedy this situation in some cases.

4.3.2 Trigger coupling

There is a separate setting for the coupling used by the triggering system of an oscilloscope. This allows the DC offset to be removed from the signal when used for triggering, while still displaying the entire waveform (including any DC offset). AC trigger coupling can be useful when triggering on a signal with a changing or unpredictable DC offset.

4.3.3 Holdoff

Holdoff is a triggering setting which ensures that a minimum amount of time elapses between triggers. The holdoff setting allows for useful triggering on a signal that satisfies the trigger condition multiple times per period, as seen in figure 4.7. In this case, the oscilloscope triggers on the shorter period sinewave that is superimposed on the larger one, causing shadows of the signal on the screen. It is likely that one would rather trigger on the longer period sinewave. Adjusting the holdoff time to slightly shorter than the larger period prevents the additional triggers, restoring a stable display.

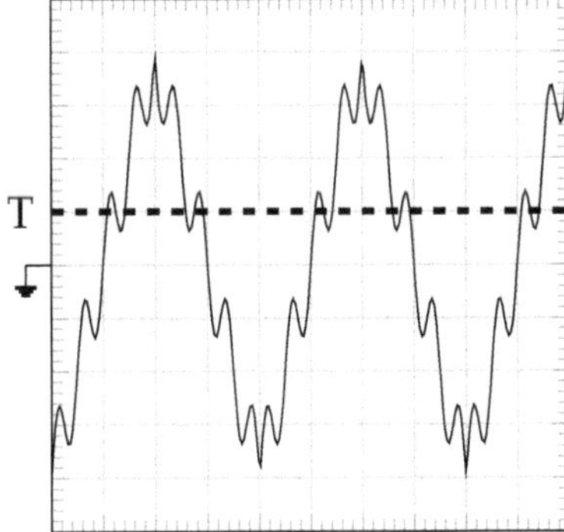

Figure 4.7: Signal that could cause multiple triggers without holdoff

Triggering of noisy signals can also benefit from an increased holdoff. Noise added to a signal may cause it to satisfy the trigger condition multiple times in quick succession. A non-zero holdoff can be used to ensure that only the first trigger event causes the oscilloscope to record data.

4.4 Frequency limitations of equipment

Every piece of equipment has a limit to the frequency range of signals it can work with. For example, a function generator has a minimum period of waveform it can output, and an oscilloscope measures signals in time at a finite speed. While oscilloscopes and function generators exist that can measure and generate signals in the tens of gigahertz, they are typically prohibitively expensive for basic lab work. Additionally, even if the function generator and oscilloscope say they are rated to a certain frequency, the interconnects between the two may limit the highest frequency that can be measured even further. It is important to always be aware of the limitations of the equipment and cabling being used. These limitations will be explored in task 4.6.5.

4.5 Prelab

Task 4.5.1: Prelab Problems

1. *Sketch* the waveform $2\sin(2\pi 10t)$ for two periods and *mark* all the rising and falling edges for a trigger level of 1 V.
2. Using the manual for your oscilloscope,
   a) *Describe* how a triggering mode that is not edge triggering works.
   b) *Describe* how to change trigger coupling from DC to AC.

Task 4.5.2: Prelab Task: Exploring oscilloscope triggering

You can complete this task using the Analog Discovery 2 or during open lab time.

1. *Set* the function generator to generate a sinusoid with the following properties:
   - Peak to peak voltage of 2 V
   - Offset of 0 V
   - Frequency of 1 kHz
2. *Connect* the output of the function generator to the oscilloscope input 1. Use autoscale to view the waveform. *Capture* a printout of the waveform.
3. Ensure that the trigger type is set to Auto. *Adjust* the trigger level and *describe* what happens. *Comment* on what happens when the trigger level is above 1 V or below −1 V. You may need to wait several seconds after adjusting the trigger level to see the effect.
4. *Set* the trigger level to 2 V. The trigger mode should still be set to auto. Vary the amplitude of the sine wave from 1 V_{p-p} to 4 V_{p-p} and *record* your observations.
5. *Set* the signal back to an amplitude of 2 V_{p-p}.
6. *Change* the trigger mode to normal and *vary* the amplitude of the sine wave from 1 V_{p-p} to 4 V_{p-p}. *Record* your observations and compare with auto triggering mode.
7. *Set* the trigger level to 0 V and change the triggering from rising edge to falling edge. Describe how this changes how the signal is plotted.

4.6 Tasks

Task 4.6.1: Input and Trigger Coupling (AD2)

1. *Set* the function generator to generate a triangle wave with the following properties:
   - Peak to peak voltage of 1 V
   - Offset of 3 V
   - Frequency of 5 kHz
2. *Use* autoscale to obtain a steady output of the waveform, then *adjust* the oscilloscope vertical scale to 1 V per division with a +2 V offset. *Capture* the waveform on the oscilloscope display using the persistence feature.
3. *Change* the **input** coupling to AC. *Adjust* the trigger level to obtain a steady waveform. *Capture* an oscilloscope screenshot showing both the AC and DC coupled signals.
4. *Set* the signal offset to 0 V. *Change* the input coupling back to DC and *set* the trigger level to 3 V. *Sweep* the offset of the signal from −3 V to 3 V. *Describe* the potential issues of the current setup if the signal to be measured has a changing offset.
5. *Change* the **triggering** coupling to AC. *Adjust* the trigger level to get a steady waveform. *Sweep* the signal offset voltage from −3 V to 3 V. *Explain* why the oscilloscope is able to maintain consistent triggering when the signal offset is changing.

Task 4.6.2: Mystery Signals (AD2)

1. *Set* the function generator to output the mystery signal provided by your instructor.
2. *Adjust* the oscilloscope so that the signal is consistently triggered on the slower waveform. *Capture* a screenshot of the oscilloscope display.
3. *Adjust* the triggering settings to trigger consistently on the faster part of the waveform. *Capture* a screenshot of the new oscilloscope display.
4. *Record* the trigger coupling, trigger level, and holdoff time you used to view the signal for each case.

Task 4.6.3: Synchronous and asynchronous signals

1. *Set* the function generator to generate a sine wave and triangle waveform with the following properties:
    - Peak to peak voltage of 8 V
    - Offset of 1 V
    - Frequency of 100 kHz
2. *Connect* the triangle wave to input 1 of the oscilloscope, and connect the sine wave to input 2.
3. *Adjust* the scaling and triggering to clearly view both waveforms using input 1 for triggering. Adjust the phase of one of the signals so that the rising edge occurs at the same time. *Capture* a screenshot of the display.
4. *Change* the frequency of the sine wave to 100 000.1 Hz. What happens to the display of the waveforms?
5. *Change* the triggering to input 2. How does the oscilloscope display change?
6. *Change* the oscilloscope to XY mode and describe the display.[a]

[a] You can plot many interesting patterns by changing the relative frequency of the signals. Look up Lissajous patterns to learn more!

Task 4.6.4: Pseudorandom Bit Sequence

1. *Set* the function generator to output a 127-bit long pseudorandom bit sequence using the instructions from your instructor.
2. *Try* to view the waveform using the oscilloscope. Are you able to view the signal?
3. The bit sequence has a period of 10 ms. In the triggering menus, *adjust* the holdoff to 9.9 ms. Why can you now view the signal? *Capture* a screenshot showing the signal.
4. *Connect* a second oscilloscope probe to the sync output of the function generator. *Change* the trigger source to be the sync signal. How does the sync signal relate to the signal? In what way does does the relationship help simplify the triggering setup? *Capture* a screenshot showing the first 10 pulses of the pseudorandom bit sequence.

Task 4.6.5: Generating very fast signals

1. *Set* the function generator to generate a square wave with the following properties:
   - Peak to peak voltage of 1.8 V
   - Offset of 0.9 V
   - Maximum frequency supported by the function generator[a].
2. *Adjust* the controls to view the generated waveform.
3. *Measure* the frequency, amplitude, and average voltage of the signal.
4. *Capture* a screenshot showing the waveform.
5. What parts of the system could cause the signal to not look like a square wave?

[a] If the maximum frequency of the oscilloscope is smaller, use that frequency.

EXPERIMENT 5

Superposition and Equivalent Circuits

5.1 Application

Nodal and mesh circuit analyses are useful to understand circuit behavior; however, it is normally not necessary to compute the voltage and current everywhere. Instead, one can transform complex circuit networks into a simpler equivalent circuits by hiding internal details. If applied correctly, the network itself changes but the original characteristics remain intact. The two heavyweights in electrical engineering are called the Norton and Thévenin equivalent circuit theorems. These theorems can take complicated and messy circuits like the grid in figure 5.1 and simplify them to a single source and a single resistor.

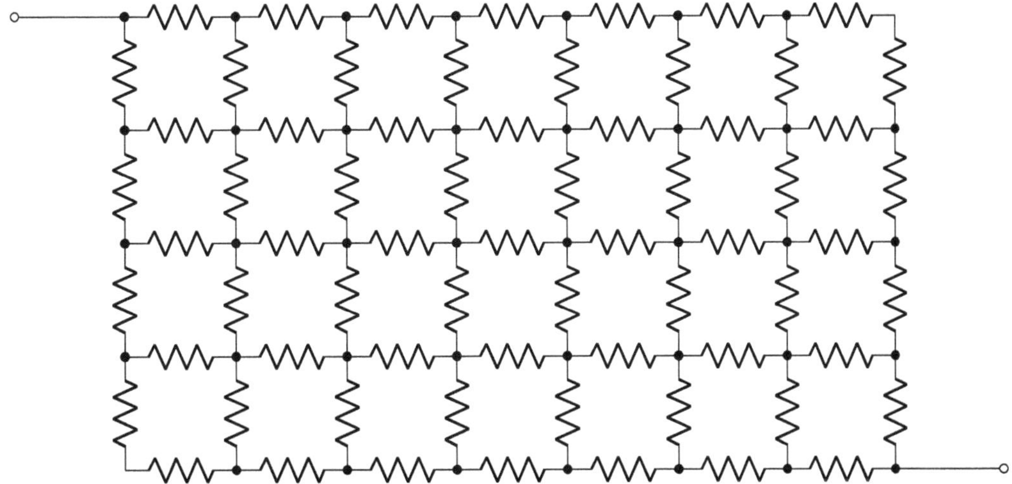

Figure 5.1: A grid of resistors that has a Thévenin equivalent of a single resistor!

5.2 Linear systems

To ease the burden of a complex problem, it is common to group system components where the input and output relationships are known. The internal details can be ignored and treated as a black box. In effect, the black box acts as its own system. It is also possible to group system boxes into yet even larger boxes. Through this process, a highly complex system can be more simply understood at a higher level.

In the words of an old and wise Professor, "it is not necessary to actually color the box black."

Figure 5.2: A graphical representation of a black box system f. It takes in input, $x(t)$, and generates output, $g(t)$. f describes the how the input will be transformed by the system.

Figure 5.2 depicts a graphical view of this approach. The characteristics of the box is denoted by the operator $f\{\cdot\}$. The input is $x(t)$ and the output $g(t)$. Mathematically, $g(t) = f\{x(t)\}$. It is important to understand that f does not necessarily indicate *how* the system accomplishes its work, but rather just describes the nature of the work. There may be many ways to physically realize the operation described by $f\{\cdot\}$.

The study of such systems is a large and broad topic. Therefore, we will focus only on one particularly important subset: linear systems. It turns out that this is not that restrictive given that engineers often approximate non-linear systems with linear ones. Making such approximations simplifies analysis but it must be done with careful attention to the conditions because the results are only valid in a certain region of operation.[1]

(1) We call inputs that allow a system to stay within these regions "small signals." You will explore this concept in great detail later in your studies.

Two necessary and sufficient properties of linear systems are scalability and additivity. That is, if a system respects both the scalability and additivity properties, then it *is* itself a linear system. The properties are routinely leveraged in signal processing, controls, communications, etc. – all engineers should understand them well!

Scalability (homogeneity)

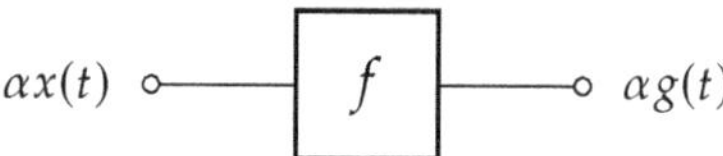

Figure 5.3: Scalability of a linear system.

Scalability of a linear system says that if the input is multiplied by the scalar α, then the output is that of the unscaled input, scaled by α. Expressed graphically in figure 5.3, the mathematical formulation is

$$f\{\alpha x(t)\} = \alpha f\{x(t)\} = \alpha g(t)$$

Additivity

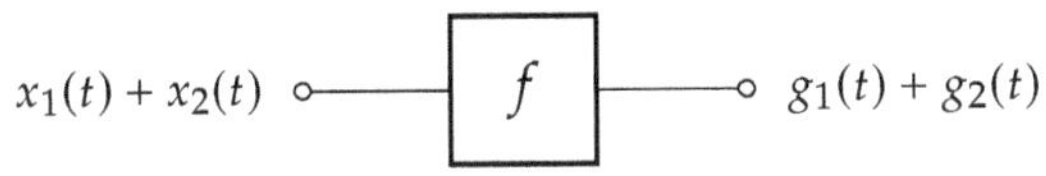

$x_1(t) + x_2(t)$ → f → $g_1(t) + g_2(t)$

Figure 5.4: Additivity of a linear system.

Additivity of a linear system says that the output is the sum of the outputs due to individual inputs. Expressed graphically in figure 5.4, the mathematical formulation is

$$\begin{aligned} f\{x_1(t) + x_2(t)\} &= f\{x_1(t)\} + f\{x_2(t)\} \\ &= g_1(t) + g_2(t) \end{aligned}$$

Notice that scalability and additivity can be combined into a more general result:

$$\begin{aligned} & f\{\alpha x_1(t) + \beta x_2(t)\} \\ &\quad = f\{\alpha x_1(t)\} + f\{\beta x_2(t)\} \\ &\qquad = \alpha g_1(t) + \beta g_2(t) \end{aligned}$$

where α and β are scalars. This leads to the superposition principal.

5.2.1 Superposition

The superposition principle(2) states that for multiple input linear systems, the output is a linear combination of the responses due to each input. Superposition is the natural result of both the additivity and scalability properties.

(2) Superposition is sometimes referred to as a property, but it is actually just a result of properties.

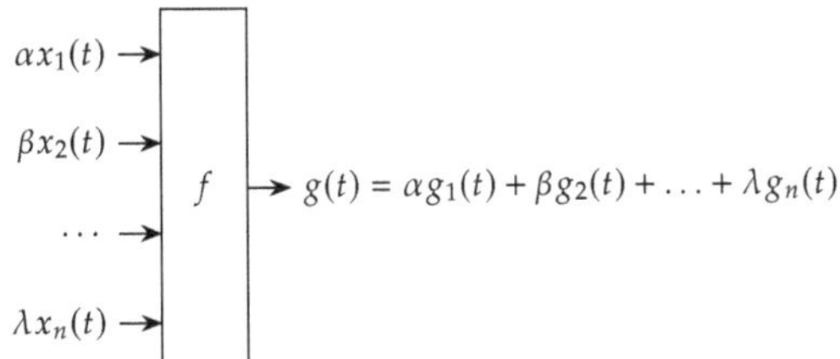

Figure 5.5: Superposition of a linear system.

To demonstrate this principle, suppose the linear system, f, has inputs $x_1(t)$, $x_2(t)$, and up to $x_n(t)$, as shown in figure 5.5. Then, the output of the system can be computed as

$$\begin{aligned} g_{\text{total}}(t) &= f\{\alpha x_1(t), \beta x_2(t), \cdots, \lambda x_n(t)\} \\ &= \alpha g_1(t) + \beta g_2(t) + \cdots + \lambda g_n(t) \end{aligned}$$

where α, β, and λ are scalars.

5.3 Circuits as a linear system

Not surprisingly circuit networks with only linear components are linear systems. Sources are the inputs and node voltages and/or currents are the outputs.

As noted before, engineers often approximate non-linear circuit devices operating at a fixed bias point with linear networks. These are called small signal equivalent circuits. Try to take note of this strategy as you learn non-linear devices, e.g., diodes and transistors.

5.3.1 Determine the linearity of a device

Linear components have linear I–V operators. For example, the voltage applied to a resistor (input) results in a current (output) where the operator is $f\{V\} = V/R$, as required by Ohm's law. This operator satisfies both the scalability and additivity requirements as shown in equation (5.1) and equation (5.2). Because both requirements are met, the operator $f\{V\}$ is linear. The same can be said if the current is taken as the input and the resulting voltage the output.

Scalability:

$$f\{\alpha V\} = \frac{\alpha V}{R} = \alpha \frac{V}{R} = \alpha f\{V\} \tag{5.1}$$

Additivity:

$$f\{V_1 + V_2\} = \frac{V_1 + V_2}{R} = \frac{V_1}{R} + \frac{V_2}{R} = f\{V_1\} + f\{V_2\} \tag{5.2}$$

5.3.2 Using linear properties to simplify circuit analysis

Superposition is particularly useful in circuit analysis. In the context of a circuit, it says that a network with multiple current or voltage sources can be reduced to several simpler networks with only one source each. The overall output is the summation of each source's contribution when the remaining sources are set to 0. The equivalent circuit to 0 A or 0 V sources are shown in figure 5.6. Setting a current source to 0 A implies an open circuit, otherwise, any voltage across it would necessitate some amount of current flow. On the other hand, setting a voltage source to 0 V implies the source must be a

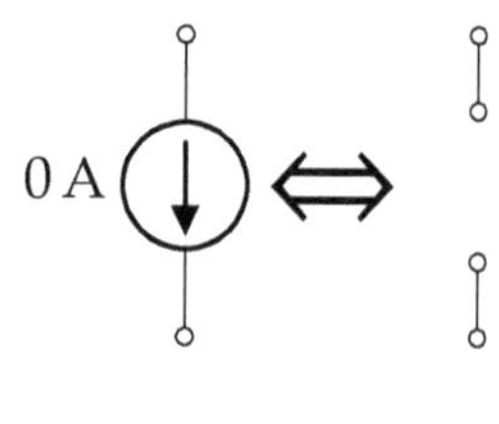

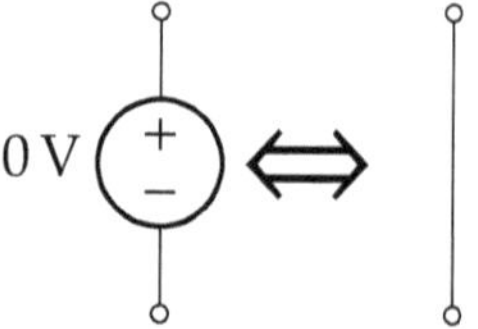

Figure 5.6: 0 V and 0 A sources

short circuit, otherwise, any current flow through it would induce a voltage. Example 5.3.1 demonstrates how to apply the superposition property to a two-source circuit.

Example 5.3.1: Superposition property of a two-source circuit

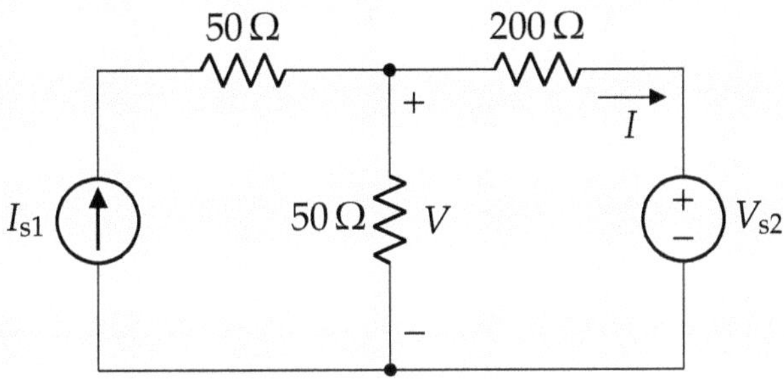

Figure 5.7: Two-source circuit for superposition analysis.

In the circuit shown in figure 5.7, the sources I_{s1} and V_{s2} are given. Determine the outputs V and I.

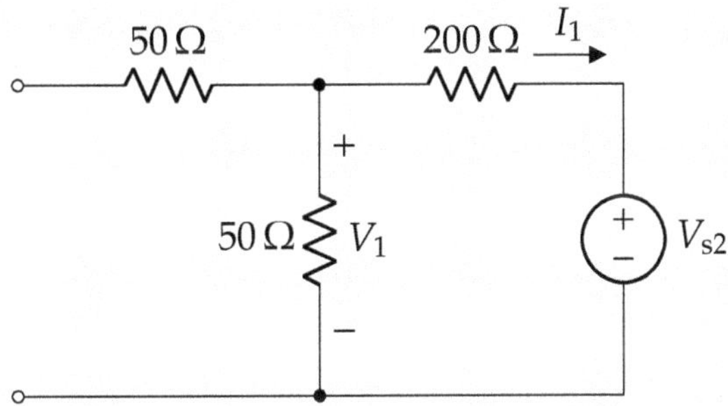

Figure 5.8: Circuit for computing contribution from V_{s2}.

To find V, apply superposition and compute the contributions from I_{s1} and V_{s2} separately. By turning off I_{s1}, the original circuit simplifies to figure 5.8. The contribution from V_{s2} is calculated using voltage division. Note that the 50 Ω plays no role because it is not part of a closed circuit.

$$V_1 = \frac{50}{200 + 50} V_{s2} = 0.2 V_{s2}$$

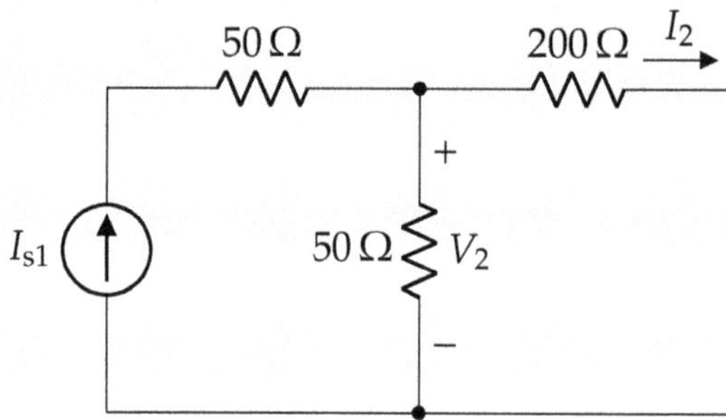

Figure 5.9: Circuit for computing contribution from I_{s1}.

To find I_{s1}'s contribution towards V, turn off V_{s2}, resulting in figure 5.9. The output contribution is

$$V_2 = 50 + \frac{50 \cdot 200}{50 + 200} I_{s1} = 40 I_{s1}$$

Therefore, the total output V is

$$V = V_1 + V_2 = 40 I_{s1} + 0.2 V_{s2}$$

Similarly, one can follow the same procedure to find I from the contributions of I_{s1} and V_{s2}

$$I = 0.2 I_{s1} - 0.004 V_{s2}$$

We encourage you to verify the above result as superposition practice.

5.4 Source transformation

Source transformation is the process of finding alternative, but equivalent circuits by converting voltage sources into current sources and vice versa. This will often expose opportunities to further combine components and simplify the equivalent network even further.

As we have previously seen, a real voltage source can be reasonably modeled with the circuit of figure 5.10[(3)]. The series resistor, R_s, accounts for the source's internal resistance and the inevitable losses suffered when practically sourcing power.

(3) Consider the function generator's equivalent circuit model from figure 2.3 on page 24.

To determine if a current source circuit could be made equivalent to a voltage source circuit, solve for each I–V relationship and compare. If the circuits have an equivalent form, then we only need to match coeffecients to determine what value current source makes an equivalent circuit.

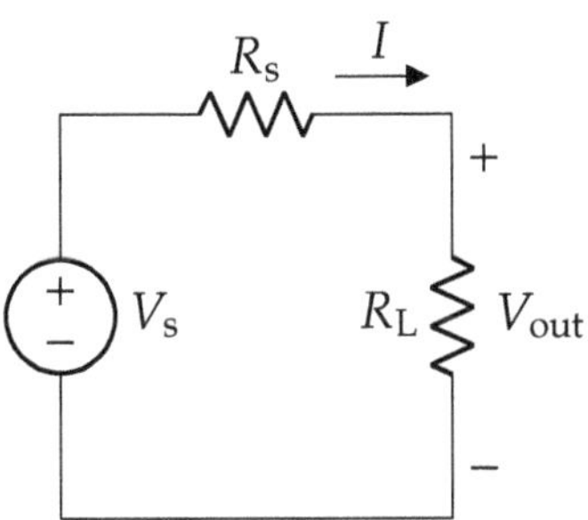

Figure 5.10: A realistic voltage source with load R_L.

By Ohm's law, the current I flowing through the load resistor, R_L, is calculated by

$$I = \frac{V_s - V_{out}}{R_s}$$
$$I = \frac{V_s}{R_s} - \frac{V_{out}}{R_s} \tag{5.3}$$

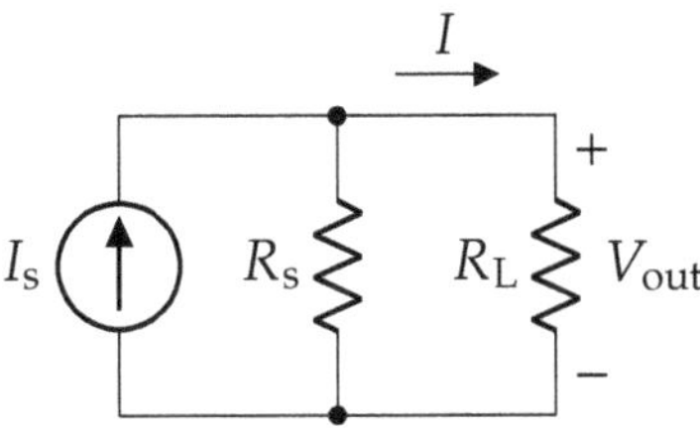

Figure 5.11: Equivalent circuit of figure 5.10 after voltage source transformation. It represents a realistic current source with internal reistance R_L connected with a load.

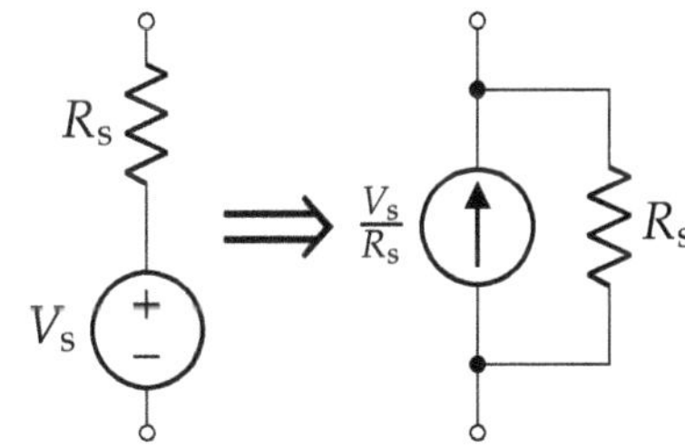

Figure 5.12: Voltage to current source transformation

From figure 5.11, the load current I can be found using KCL at either node.

$$I_s = I + \frac{V_{out}}{R_s}$$
$$I = I_s - \frac{V_{out}}{R_s} \tag{5.4}$$

As stated before, Figures 5.10 and 5.11 are equivalent if they have the same I–V characteristics. Comparing equations (5.3) and (5.4), it is clear that equivalence occurs when $I_s = \frac{V_s}{R_s}$. Therefore, a voltage source with voltage V_s and a series resistance R_s is equivalent to a current source with current $\frac{V_s}{R_s}$ in parallel with the same resistance R_s. Similarly, a current source with current I_s and a parallel resistance R_s is equivalent to a voltage source with voltage $I_s \times R_s$ in series with the same resistance R_s.

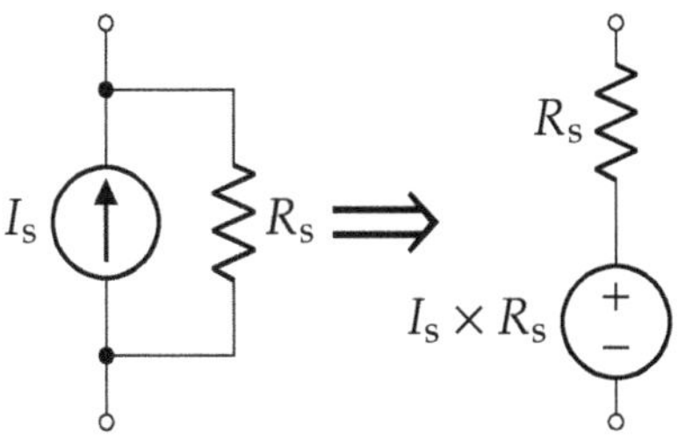

Figure 5.13: Current to voltage source transformation

5.5 Thévenin and Norton theorems

Taking the previous results to their logical conclusion ends up at the Thévenin and Norton theorems. These theorems are another key result of linearity and are very important tools for analyzing circuits. They play a large role in predicting the side effects of connecting two circuits together. Example 5.5.1 demonstrates using these methods to find an equivalent circuit model.

Thévenin Theorem

The Thévenin theorem states that a particular node in a linear circuit has an equivalent circuit which contains only one independent voltage source in series with a resistor. The independent voltage source is denoted V_{oc}, and the resistor, R_{th}, is referred to as the Thévenin resistance. Figure 5.14 shows the basic form of a Thévenin circuit model.

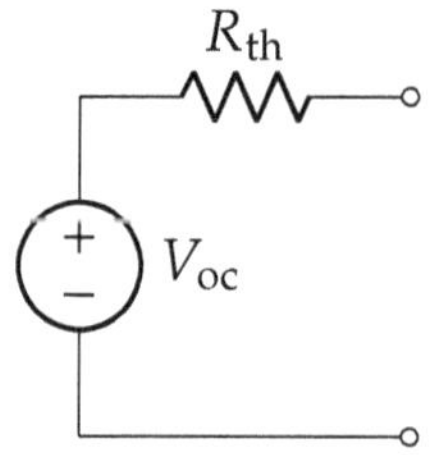

Figure 5.14: Thévenin circuit model.

Norton Theorem

Norton's theorem states that a particular linear network has an equivalent circuit which contains only one independent current source in parallel with a resistor. The independent current source is denoted as I_{sc}, and the value is called the short circuit current. The resistor, R_n, is referred to as the Norton equivalent resistance. It is important to note that $R_n = R_{th}$ because a Norton circuit is just a source transform of a Thévenin circuit. Figure 5.15 shows the basic form of a Norton circuit model.

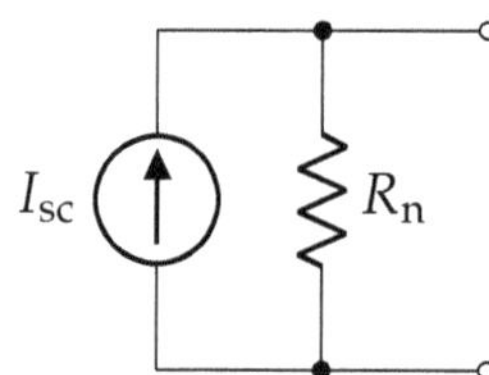

Figure 5.15: Norton circuit model.

Example 5.5.1: Find the Thévenin and Norton equivalent circuit seen by R_L.

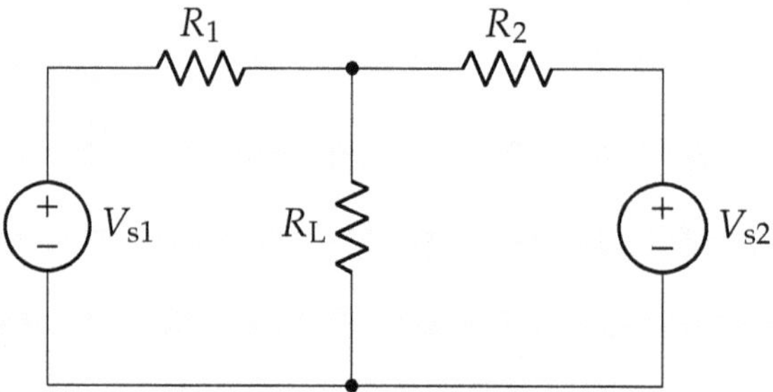

Figure 5.16: Circuit for finding Thévenin and Norton equivalent.

In figure 5.16, $R_1 = 50\,\Omega$, $R_2 = 200\,\Omega$, $R_L = 50\,\Omega$, $V_{s1} = 15\,\text{V}$, and $V_{s2} = 10\,\text{V}$. The notation "seen by R_L" means the equivalent circuit of interest is the one where R_L's nodes are the output.

To find the Thévenin equivalent circuit at that point, we start by removing the load resistor and deactivating the voltage sources to obtain the circuit shown in figure 5.17. From this we can simplify the network's resistance to a single resistor and call it the Thévenin resistance.

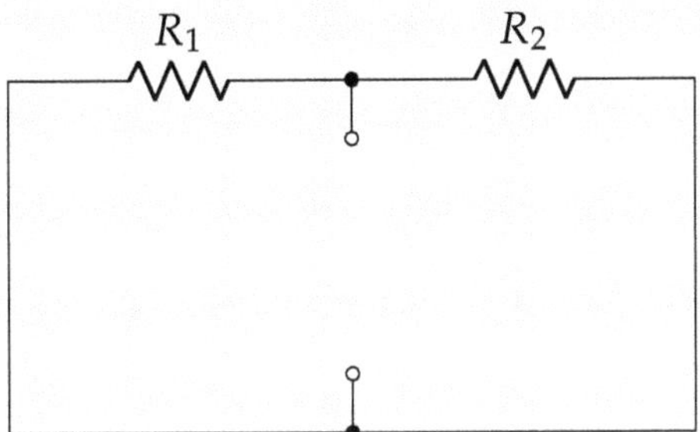

Figure 5.17: Original circuit with shorted voltage sources and removal of load resistor.

Therefore, the equivalent resistance R_{th} is

$$\begin{aligned} R_{th} &= R_1 \parallel R_2 \\ &= \frac{R_1 R_2}{R_1 + R_2} = \frac{50 \cdot 200}{50 + 200} \\ &= 40\,\Omega \end{aligned}$$

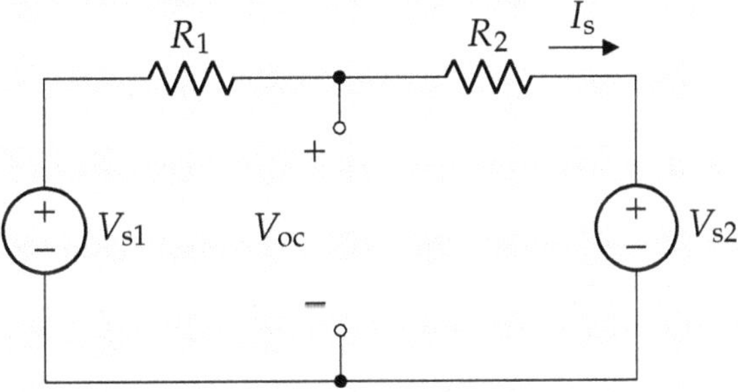

Figure 5.18: Original circuit without the load resistor.

The Thévenin voltage is found by reactivating the sources as in figure 5.18 and computing the voltage at the output. The current through the resistors is

$$\begin{aligned} I_s &= \frac{V_{s1} - V_{s2}}{R_1 + R_2} \\ &= \frac{15 - 10}{200 + 50} = \frac{5}{250} = 0.02\,\text{A} \end{aligned}$$

Therefore, the equivalent voltage is

$$V_{oc} = 10 + 200 \cdot 0.02 = 14\,\text{V}$$

Putting the two pieces together results in the Thévenin equivalent circuit shown in figure 5.19. The simplified circuit and figure 5.16 behave identically from an I–V curve perspective.

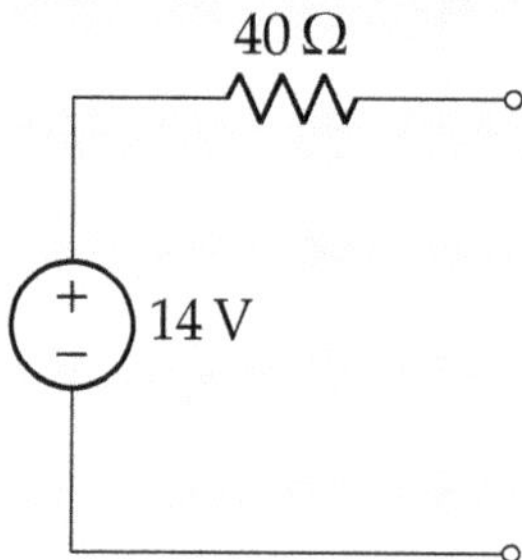

Figure 5.19: The Thévenin equivalent of the original circuit.

Instead of finding the Norton equivalent circuit directly, it can simply be found by source transformation.

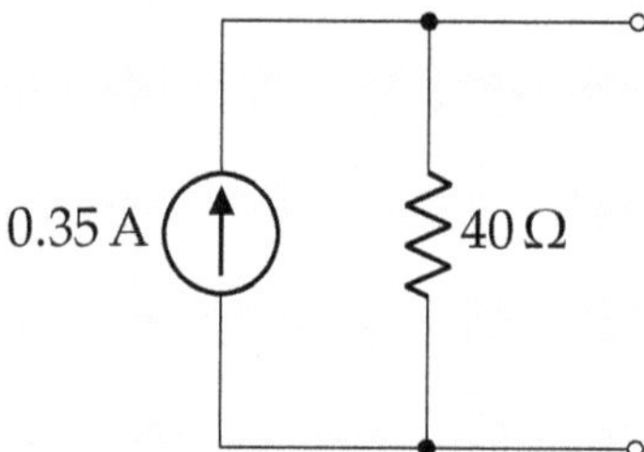

Figure 5.20: The Norton equivalent circuit.

Challenge: Try directly computing the Norton equivalent, and then find the Thévenin model using source transformations.

5.6 Implications of equivalent circuits

Thévenin and Norton equivalent circuits are not only used to simplify complex circuits for mathematical analysis. They also provide real intuition for practical applications.

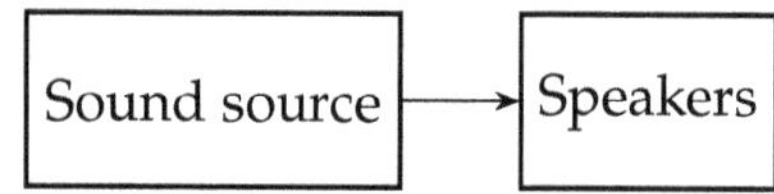

Figure 5.21: Block diagram of a sound source driving a speaker.

Figure 5.21 shows a high-level block diagram for the connection of a sound source and speaker. To start building an understanding of how the connection will affect the whole,(4) replace each block with equivalent Thévenin circuit models, as shown in figure 5.22. In is important to recognize that the actual source and speakers circuits are likely much more complicated. These are *equivalent* circuits with the same outputs for the same inputs as the original. These circuit analogies are easier to work with and effect each other in the same way the actual circuits do.

(4) And it *will* affect it!

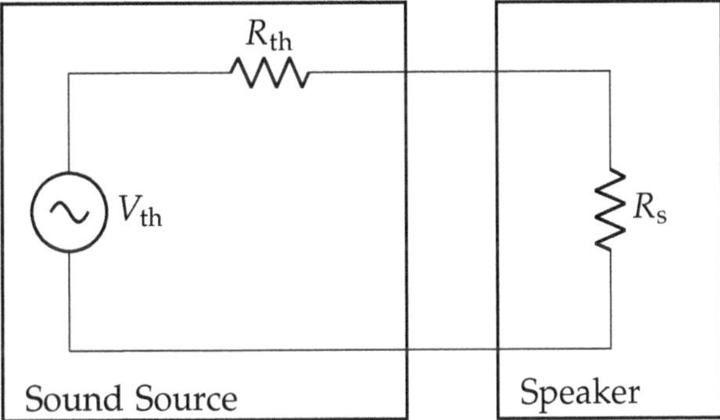

Figure 5.22: Circuit model of a source/speaker system using Thévenin equivalent circuits for both the source and the speaker.

Let us assume this is a loud speaker and the objective is to deliver as much of the source's signal to the speaker as possible. However, the equivalent circuit suggests that loading may be an issue if $R_{th} \gg R_s$, as we saw to be the case in example 2.4.1. You can see the effects of loading by considering how the two resistors relate to a voltage divider.

5.6.1 Voltage buffer

One way of mitigating the loading caused by a large source resistance is to increase the amplitude of the source; however, the source may reach its limit before the desired voltage is achieved. Additionally, even if the desired output voltage could be reached, the source amplitude would be load dependent, requiring continuous tuning to maintain the desired output. Finally, a significant power would be wastefully(5) consumed by the source's internal resistor (R_{th}).

(5) It's not wasteful if you are trying to build a space heater as well! Resistors dissipate all of their energy as heat.

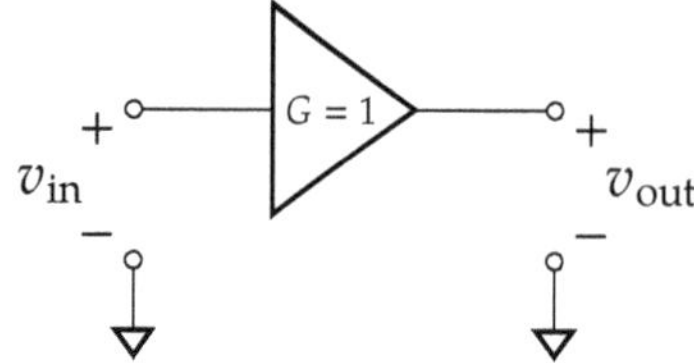

Figure 5.23: Schematic symbol for a buffer circuit, where $G = 1$ indicates the voltage gain is one.

A better solution would be to deploy a voltage buffer[(6)] to *transform* the source resistance of the function generator to a smaller one. A commonly used schematic symbol for such a buffer is shown in figure 5.23. The details of a voltage buffer are beyond the scope of this experiment, but, in general, a buffer provides at the following benefits:

(6) Sometimes called a voltage follower or unity gain amplifier.

1. A low output resistance derived from external power.
2. A large input resistance – often times so large that one cannot easily measure it.
3. A voltage gain of 1 – the input signal equals the output.[(7)]

(7) To be fair, there are plenty of amplifiers with gain > 1 that also provide benefit 1 and 2. In practice, the term "buffer" implies a gain of 1, but often an amplifier with gain > 1 can replace the need for a buffer.

Circuit model

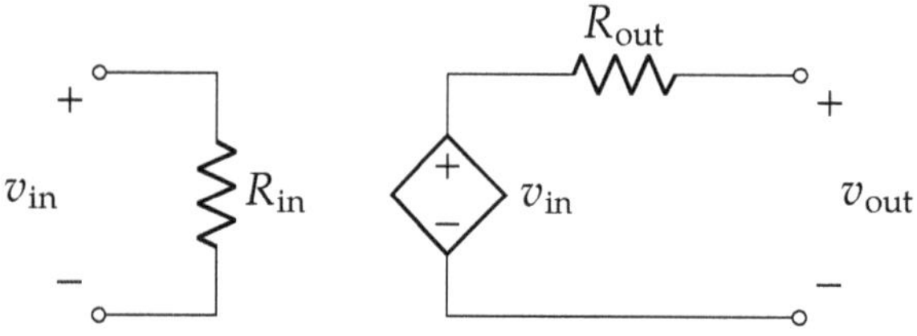

Figure 5.24: Linear circuit model of an ideal voltage buffer

To enable simple linear circuit analysis, the buffer can be replaced with a reasonable equivalent circuit shown in figure 5.24, sometimes called a Voltage-Controlled Voltage-Source (VCVS). It can, among other methods, be realized with an op amp. As already noted, a decent voltage buffer will have $R_{in} \gg R_{out}$.[(8)]

(8) Consider: what does that imply about the apparent voltage divider formed between the two circuits?

Notice that the voltage of the dependent source is equal to the voltage dropped across the buffer's input resistance. Therefore, if the voltage buffer is under-specified and still loads the original source, the loss will be preserved and seen at the output.

Example 5.6.1: Avoiding loading with a buffer

In this example, we consider the benefit of adding a voltage buffer to the output of a function generator when the load resistance is small. Figure 5.25 demonstrates this scenario schematically.

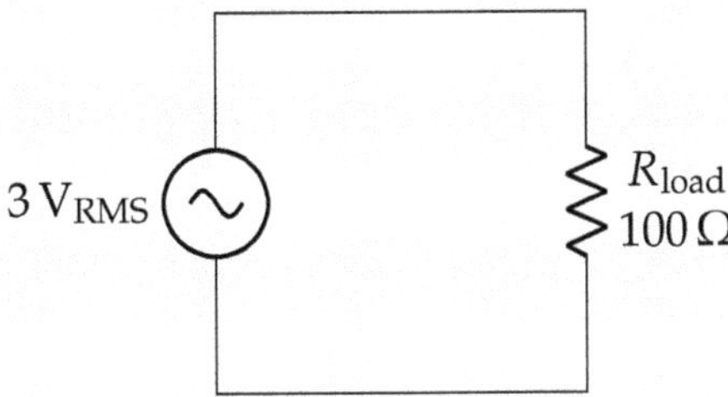

Figure 5.25: Heavily loaded function generator with "real" sources.

To better see that the source is heavily loaded, we replaced the real source with an equivalent in figure 5.26.

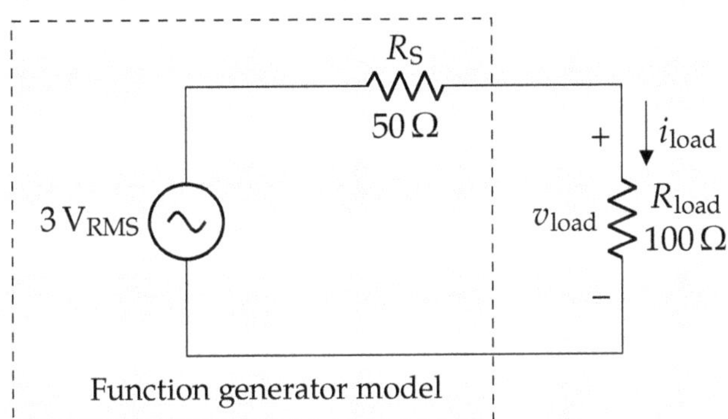

Figure 5.26: Same as figure 5.25, but with equivalent model for the function generator. The source is ideal.

Notice that the source resistance is on the same order as the load. Therefore, significant loading will occur, and the voltage delivered is just

$$v_{\text{load}} = 3\,\text{V}_{\text{RMS}} \times \frac{R_{\text{load}}}{R_{\text{S}} + R_{\text{load}}} = 3\,\text{V}_{\text{RMS}} \times \frac{100\,\Omega}{150\,\Omega} = 2\,\text{V}_{\text{RMS}}.$$

That is 33% less than expected!

We can utilize a voltage buffer to recover the original amplitude of the input signal – assuming the buffer itself is capable of delivering the load's power draw (R_{out} is small enough). The nominal circuit is shown in figure 5.27.

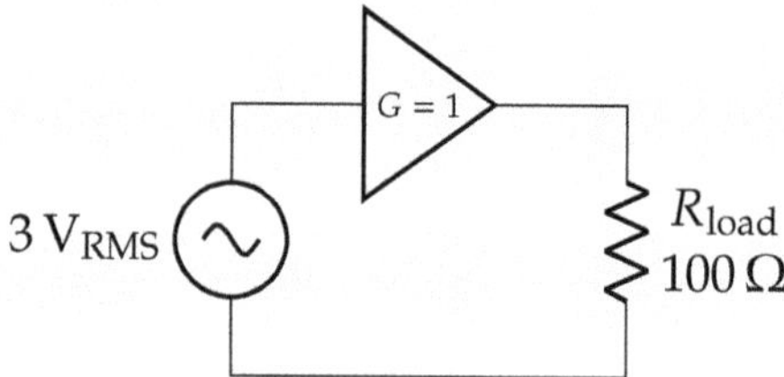

Figure 5.27: A buffered function generator to address loading issues.

For the purposes of this example, we assume the voltage buffer has a modest $R_{\text{in}} = 500\,\text{k}\Omega$ and $R_{\text{out}} = 2\,\Omega$. These are parameters that one would either derive from a buffer circuit design, measure, and/or lookup in a datasheet.

To move the analysis forward we redraw the circuit. One should replace each real and/or non-linear element with a reasonable linear equivalent, as done in figure 5.28. Of course, "reasonable" is hard to define and may not be clear at this point in time. With practice, your intuition will begin to help you understand when equivalent circuit models can be used.

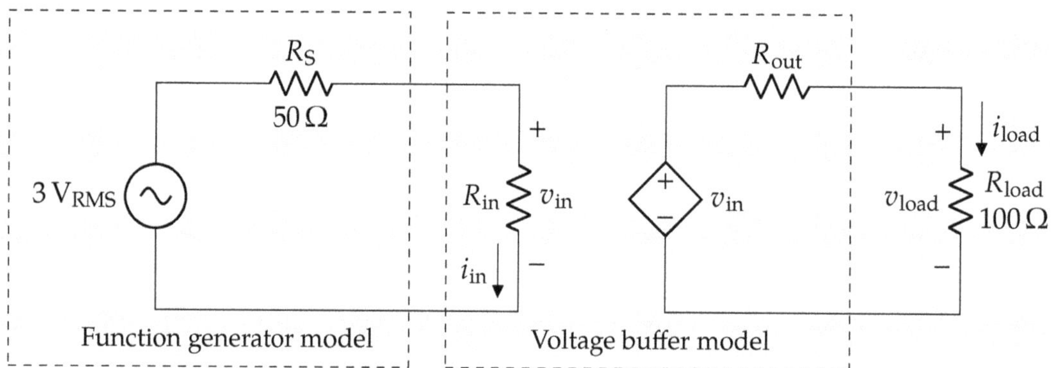

Figure 5.28: Loaded function generator and voltage buffer from figure 5.27 replaced with linear models.

To find the output, solve for the value of the dependent source. This is really the same analysis as before, expect now R_{in} is much larger than R_{S}.

$$v_{\text{in}} = 3\,\text{V}_{\text{RMS}} \times \frac{R_{\text{in}}}{R_{\text{S}} + R_{\text{in}}} = 3\,\text{V}_{\text{RMS}} \times \frac{500\,\text{k}\Omega}{500.05\,\text{k}\Omega} = 2.99\,\text{V}_{\text{RMS}}$$

Repeating the same type of calculation with the dependent source output, we have

$$v_{\text{load}} = 2.99\,\text{V}_{\text{RMS}} \times \frac{R_{\text{load}}}{R_{\text{out}} + R_{\text{load}}} = 3\,\text{V}_{\text{RMS}} \times \frac{100\,\Omega}{102\,\Omega} = 2.93\,\text{V}_{\text{RMS}}.$$

Now v_{load} is only 2.2% less than expected!

Modern buffers or amplifiers typically have much greater input and smaller output resistances than this example, only further improving upon the results seen here.

5.7 The integrated circuit (IC)

Shifting gears for a moment, we must to stop to talk about integrated circuits (ICs) because you will use them in this experiment. ICs are devices that contain a combination of circuit components typically integrated on a piece of silicon. They are ubiquitous in any electronic item. The small black packages primarily contain transistors that act as electronically controllable switches. An analog device may contain hundreds of transistors in a single package, and a modern processor contains *billions*.(9) Conveniently, these complicated circuits can often be modeled with equivalent linear circuits.

(9) No need to fret, ICs with billions of transistors are mostly digital circuits that are realized automatically by computer. Specialized hardware description languages (VHDL, Verilog, etc.) capture the logic and are compiled to schematics and more.

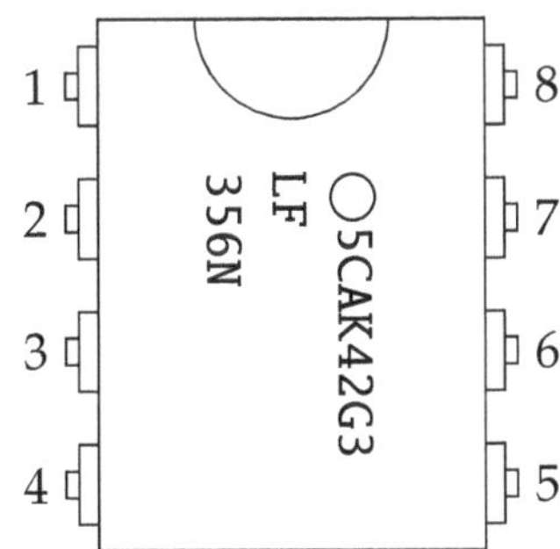

Figure 5.29: An 8 pin DIP package

In this experiment, you will use an op amp built into a dual inline package (DIP). Dual inline refers to the fact that there are two rows of pins in line with each other. The package conveniently fits across the center divot in a breadboard, making prototyping simple. Most electronic packages are labeled counter clockwise with pin 1 in the top left corner. The "top" of a package should be labeled with a dot or a divot in the plastic case. An example DIP package is shown in figure 5.29.

Figure 5.30: ICs come in many shapes and sizes!

The text on the package will usually have the manufacturer logo, a date code indicating time of manufacture, and the actual part number. For the example in figure 5.29, the circle is the manufacturer logo, `5CAK42G3` is the date code, and `LF356` is the part number. While there is plenty of room for a part number on a DIP package, smaller parts have shortened versions and can be difficult to identify without the corresponding datasheet.

5.8 Prelab

Task 5.8.1: Prelab questions

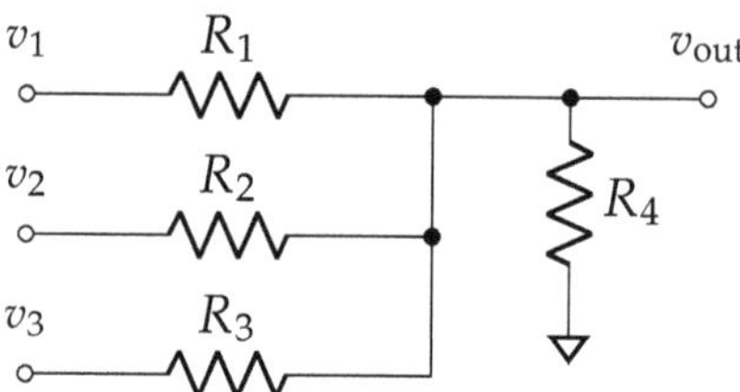

Figure 5.31: Circuit for observing its superposition property.

1. Using hand calculations and superposition, *compute* the expected output voltage v_{out} when $R_1 = R_2 = 10\,\text{k}\Omega$, $R_3 = R_4 = 80\,\text{k}\Omega$ and
   - v_1 is a DC voltage of 2.25 V,
   - v_2 is a triangle wave with an amplitude of 4.5 V at 1 kHz,
   - and v_3 is a sine wave with an amplitude of 3.6 V at 10 kHz.
2. *Plot* the expected output of prelab question 1 using Python's NumPy, MATLAB® or similar software
3. *Locate* the datasheet for the JFET-input operational amplifier (LF356N) [1]. *Copy* the package pinout and *lookup* the specified R_{in}.
4. *Complete* the SPICE tutorial on the course website. *Generate* schematics and simulations for both the DC and transient simulation.

Task 5.8.2: Prelab task: Verification of the superposition principal

You can complete this task with the Analog Discovery 2.

1. *Build* the circuit in figure 5.31 on your breadboard.
2. For each input v_1, v_2, and v_3,
   a) *Set* the other inputs to 0 V
   b) *Measure* the peak to peak and average voltage of v_{OUT} using the automatic measurement tools.
   c) *Capture* an oscilloscope screenshot showing v_{IN} and v_{OUT}.
   d) *Calculate* the percent error for each measurement.
3. *Apply* all three signals simultaneously and *capture* a plot of the input and output waveforms using the oscilloscope. Does the plot match your the results from prelab question 2?
4. *Describe* how these results consistent with the superposition principal.

5.9 Tasks

Task 5.9.1: Loading with a real signal

For this task, you will use a practical voltage buffer to measure the effects of loading. The gory details of operational amplifiers (op amps) are out of the scope of this experiment. For now, one just needs to understand that the circuit in figure 5.32 implements a voltage buffer, where the v_{in} signal is duplicated and driven out at v_{out}. Please use the LF356N op amp. You can find the device's pinout in the datasheet.

The capacitors in this circuit are called decoupling capacitors and are important to the correct operation of an IC. You should **always** include decoupling capacitors whenever using ICs. In this case, 0.1 µF is sufficient.

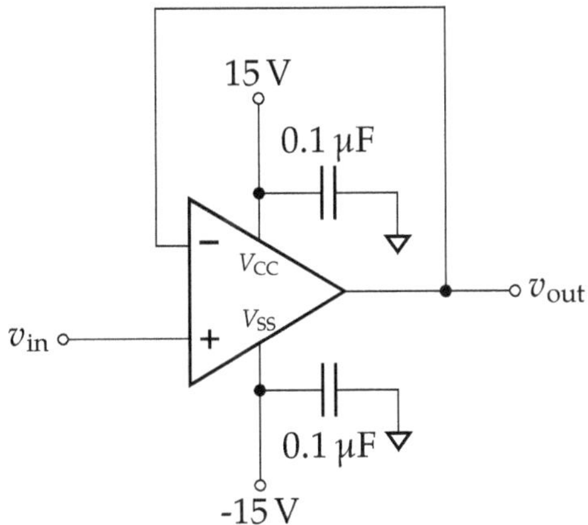

Figure 5.32: An op amp based voltage buffer.

1. *Connect* a function generator to the speaker and *play* a 250 mV $\sin(2\pi \times 2000t)$ signal.
2. *Measure* the RMS voltage across the speaker. Is it what you expected? Why or why not? *Compute* a percent error. Keep in the mind the relative loudness of the sound.
3. *Connect* the op amp voltage buffer between the function generator and the speaker.
4. *Measure* the RMS voltage across the speaker and the RMS voltage at the input of the buffer. Are they what you expected? Why or why not? *Compute* a percent error.
5. Is the sound louder than without the buffer?
6. *Estimate* the Thévenin equivalent resistance of the buffer output using the amplitude of the signal across the speaker.[a]

[a] Hint: use the voltage division equation. You can solve for R_{th} using v_{OUT}, v_{IN}, and R_L.

Task 5.9.2: Thévenin equivalent circuit design

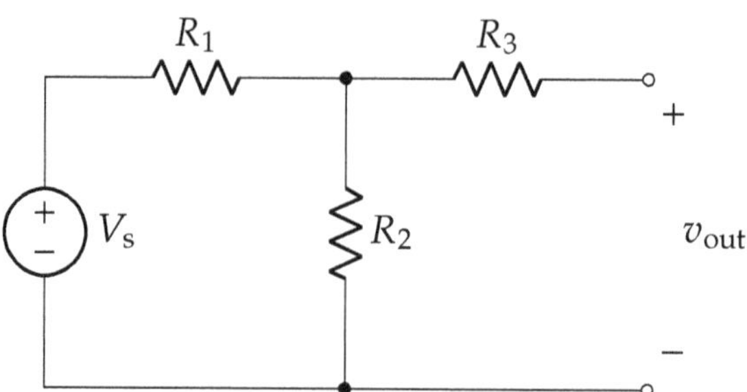

Figure 5.33: Thévenin equivalent circuit design.

1. *Design* the circuit in figure 5.33 to have a Thévenin equivalent voltage of 2.5 V and a Thévenin equivalent resistance of 1.5 kΩ. *Record* V_s, R_1, R_2 and R_3 for your design.
2. *Develop* a method for measuring the Thévenin equivalent circuit for a two terminal network.
3. *Apply* your method to verify the design.
4. *Draw* the measured Thévenin equivalent circuit. Does this agree with what you expected?
5. **Extra Credit:** *Design* a test setup and *write* an SCPI program to automate finding a Thévenin equivalent circuit.
   - *Include* diagrams of your test setup and all code written
   - *Use* your design to test the circuit in figure 5.33.
   - Does your design work for circuits without any sources? If not, how could you modify it to work for source-free circuits?

Task 5.9.3: Simulate the superposition principal with SPICE

1. *Create* a SPICE simulation for the circuit in figure 5.31.
2. *Simulate* and *plot* the individual v_{out} response due to each input v_1, v_2 and v_3, as defined in prelab question 1.[a] Do the plots match your prelab results?
3. *Obtain* the output waveform v_{out} for all three inputs applied simultaneously. Does the plot match your prelab results?
4. *Explain* how these results are or are not consistent with the superposition principal.

[a] Remember to set the sources not in use to 0 V!

EXPERIMENT 6

Lab Practical 1

Note: The gray boxes like this indicate values that will be given to you at the time of the practical. To practice, try coming up with your own.

6.1 Take home exam portion (25)

Complete this section *before* your in-lab practical exam.

1. Draw the circuit model for each piece of test equipment. Label all components.

   a) **(1)** Power supply unit

   b) **(1)** Function generator

   c) **(1)** Ammeter

   d) **(1)** Voltmeter

   e) **(1)** Ohmmeter

   f) **(2)** Oscilloscope
   (Indicate both AC and DC coupling)

2. **(6)** Study the signals shown in figure 6.1 and complete the blanks (including units) using the voltage and time scale given by your instructor.

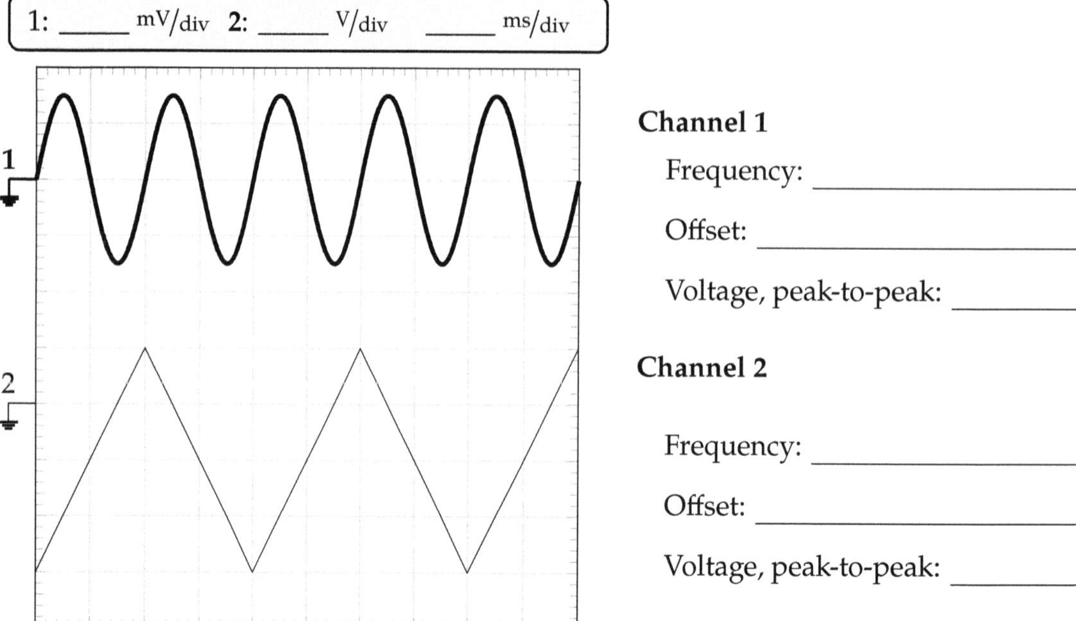

Figure 6.1: Oscilloscope capture of signals.

3. **(6)** Plot the signal provided by your instructor on the oscilloscope screen shown in figure 6.2. Completely annotate the figure and use only multiples of 10 for voltage and time per division.

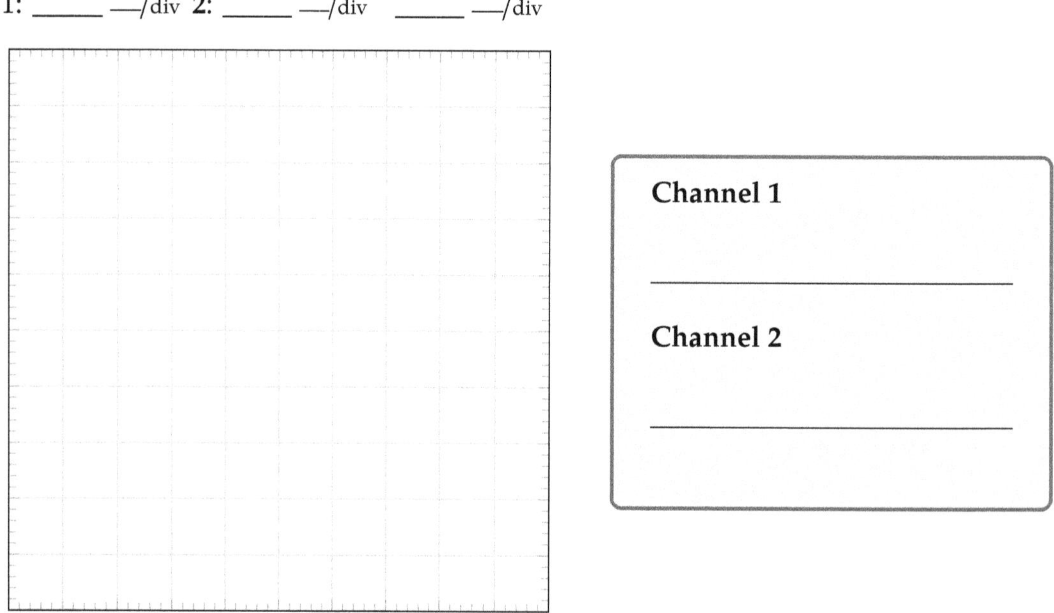

Figure 6.2: Blank oscilloscope screen capture.

Desired thévenin equivalent parameters

r_{th} = __________ Ω v_{oc} = __________ V

note: $v_{oc}/r_{th} < 100\,\text{mA}$, $r_{th} < 2.5\,\text{k}\Omega$, and $v_{oc} < 10\,\text{V}$.

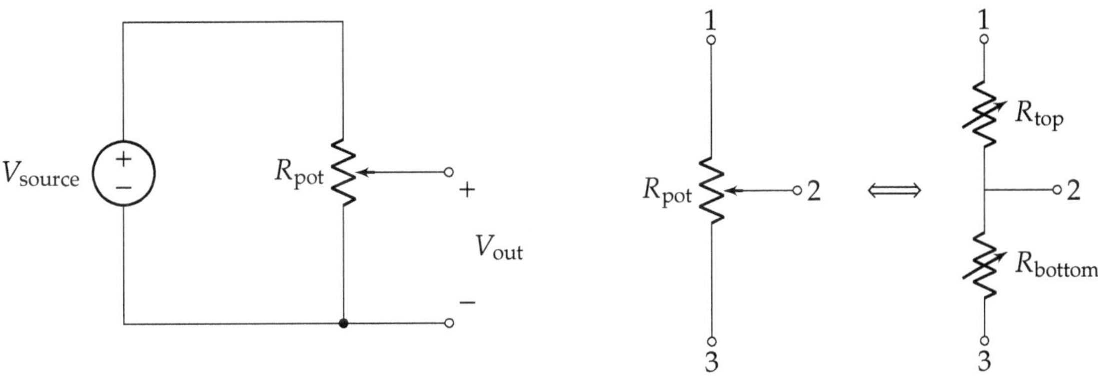

Figure 6.3: Circuit to design for a specific Thévenin equivalent. The potentiometer model is shown in figure 6.4.

Figure 6.4: A potentiometer is two series adjustable resistors constrained by $R_{top} + R_{bottom}$ = constant.

4. **(5)** Using the circuit from figure 6.3 and $R_{pot} = R_{top} + R_{bottom} = 10\,\text{k}\Omega$, compute R_{top}, R_{bottom}, and V_{source} such that the desired Thévenin equivalent circuit is obtained.. Show all work.

5. **(1)** What is the Norton equivalent of the circuit designed in problem 4?

6.2 In-lab exam portion (75)

Load Resistance: R_{load} = ________ Ω
note: R_{load} will be a value in your kit.

Task 6.2.1: Verify the equivalent circuit from problem 4 via simulation.

1. **(10)** *Create* a SPICE simulation of figure 6.3 using the values calculated in problem 4 and the circuit model in figure 6.4.
2. **(5)** *Run* an operating point analysis to *measure* V_{oc} = ______ V.
3. **(5)** *Run* an operating point analysis to *measure*: I_{sc} = ______ mA.

Checkpoint. You may continue to work on other sections, but do not close the simulation results until they have been verified by your instructor.

Verified: __________ Instructor Notes:

Task 6.2.2: Build the Thévenin equivalent circuit from problem 4.

1. **(1)** *Set* the current limit to 100 mA.
2. **(10)** *Build* the circuit from figure 6.3. *Adjust* the potentiometer to the values calculated in problem 4.
3. **(5)** *Measure*: V_{oc} = ______ V.
4. **(5)** *Measure*: I_{sc} = ______ mA.
5. **(2)** *Measure* the actual resistance of R_{load}. $R_{\text{load,meas}}$ = ______ Ω
6. *Attach* R_{load} from V_{out} to ground.
7. **(2)** *Measure* $V_{\text{load,meas}}$ = ______ V

Checkpoint. You may continue to work on other sections, but do not disassemble your work until it has been verified by your instructor.

Verified: __________ Instructor Notes:

6. **(4)** Compute P_{load} using $V_{\text{load,meas}}$ and $R_{\text{load,meas}}$, then calculate the percent error from the expected P_{load}.

Desired equivalent resistance

$r_{\text{eq}} =$ ________ $\Omega \pm 5\%$

Your instructor will provide the connections of the resistors below.

7. **(5)** Pick R_1 through R_5 in figure 6.5 such that the network has an equivalent resistance in the given range for r_{eq}. You can only use up to $3 \times 1\,\text{k}\Omega$, $2 \times 10\,\text{k}\Omega$, and $2 \times 100\,\text{k}\Omega$ resistors.

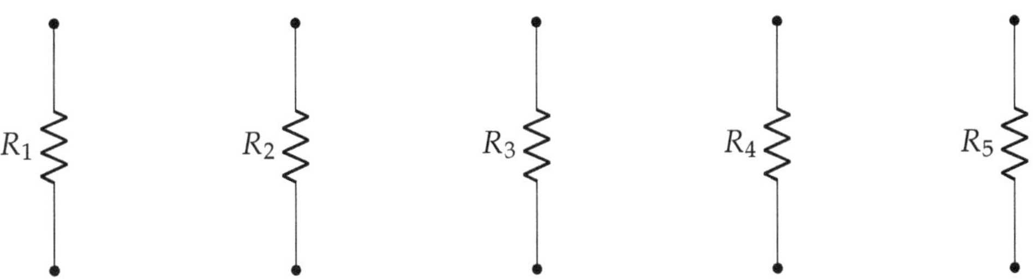

Figure 6.5: An array of resistors

Task 6.2.3: Verify the equivalent circuit from problem 7

1. **(5)** Construct the circuit designed in figure 6.5 on your breadboard.
2. **(2)** Measure the resistance of the network: $r_{\text{eq}} =$ ______ Ω.
3. **(2)** Compute the percent error of the constructed network: ______ %

Checkpoint. You may continue to work on other sections, but do not dissasemble the circuit until it has been verified by your instructor.

Verified: __________ Instructor Notes:

Task 6.2.4: Oscilloscope skills

1. **(5)** *Recreate* the oscilloscope plot from problem 3 using your function generator and oscilloscope. *Capture* a screen shot.
2. **(5)** *Run* the computer software to generate two mystery signals on the function generator. *Adjust* the triggering settings until the display is stable. *Capture* the oscilloscope display. You can trigger off of any signal. You may need to adjust the trigger level and holdoff.
3. **(2)** *Capture* an oscilloscope screen shot in XY mode using the same mystery signals from step 2.

Checkpoint. You may continue to work on other sections, but do not disassemble your work until it has been verified by your instructor.

Verified: __________ Instructor Notes:

EXPERIMENT 7

Inductors and Capacitors

7.1 Application

Capacitors and inductors are essential to the operation of analog filters and power electronics. Capacitors are frequently used to store charge in case of a sudden inrush of current draw. Inductors can be used to filter out high frequency noise in a power cable and are also an important part of high efficiency power supply circuits.

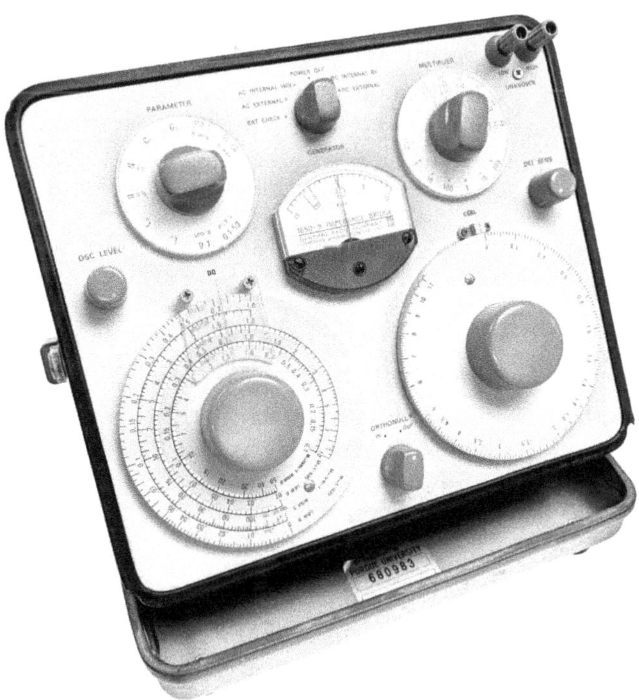

Figure 7.1: An impedance bridge used for measuring inductance, capacitance, and resistance. The 1650B was originally designed and built in the late 1960's and is about 20 times the size of the LCR meter in figure 7.8.

7.2 The capacitor

A capacitor consists of two conductors separated by a dielectric medium. When there is a voltage applied to the two conductors, charges build up on the conductors and the electric field between them increases. The amount of charge (Q) stored by a capacitor is proportional to the voltage (V_C) with a constant of capacitance (C) according to equation (7.1). Capacitance is measured in a unit called Farads (F).

Figure 7.2: An assortment of capacitors

$$Q = CV_C \tag{7.1}$$

By taking the derivative of both sides of equation (7.1), we can obtain an equation that relates the current to the derivative of the voltage.(1)

(1) Recall that $\frac{dQ(t)}{dt} = I_C(t)$

$$i_C(t) = C\frac{dv_C(t)}{dt} \tag{7.2}$$

7.2.1 Parallel plate capacitors

The simplest capacitor to analyze is the parallel plate capacitor. It consists of two conducting plates separated by a constant distance filled with a dielectric medium, as seen in figure 7.3. The capacitance of a parallel plate capacitor is given in equation (7.3). ϵ_0 is the permittivity of free space ($\epsilon_0 \approx 8.85 \times 10^{-12}\,\mathrm{F\,m^{-1}}$), ϵ_r is the relative permittivity of the dielectric, A is the area of the plates, and d is the distance between the plates. This equation works for plates of any shape.

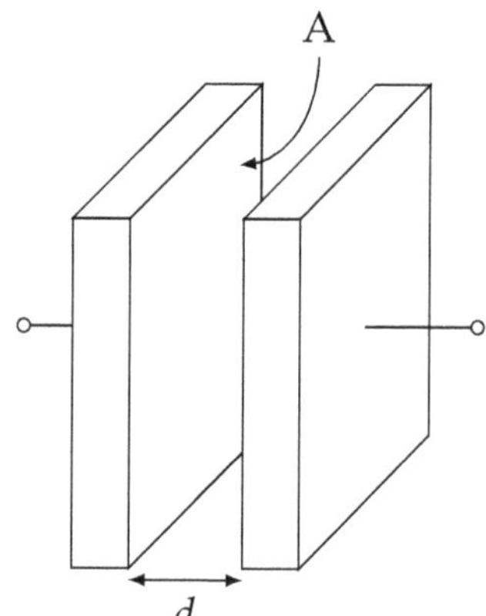

Figure 7.3: A Parallel Plate Capacitor

$$C = \frac{\epsilon_0 \epsilon_r A}{d} \tag{7.3}$$

ϵ_r is approximately 1 for air. Commercially built capacitors are made with ceramic or oxide layers for their dielectrics. These materials have a much higher dielectric constant and, therefore, much higher capacitance per unit area.

7.3 The inductor

$$v_L(t) = L\frac{di_L(t)}{dt} \tag{7.4}$$

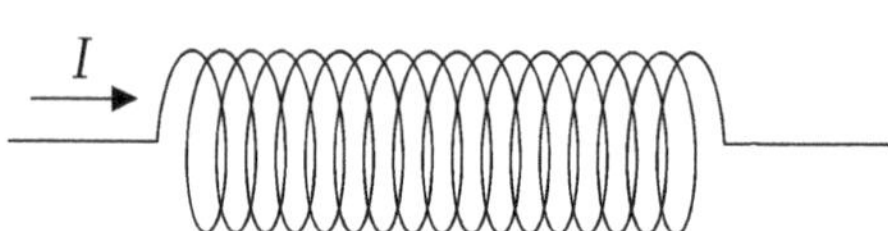

Figure 7.4: The shape of a typical inductor

When current flows through any wire, it generates a magnetic field around the wire. When coiled tightly together, the magnetic fields add up to become relatively large. These magnetic fields are able to store energy that can be released at a later time. For circuit analysis, inductors operate according to equation (7.4). The constant L is called inductance and is measured in Henrys (H).

7.3.1 Coil inductors

The most simple inductor is a wire wrapped into the shape of a spring, as shown in figure 7.4. The inductance of a coil in this shape is given in equation (7.5), where N is the number of turns, μ_0 is the permeability of free space ($\mu_0 \approx 4\pi \times 10^{-7}\mathrm{H\,m^{-1}}$), μ_r is the relative permeability of the material in the center of the inductor (usually air or metal), R is the radius of the coil, and r is the radius of the wire.

Figure 7.5: An assortment of inductors

$$L_{\text{coil}} = N^2\mu_0\mu_r R\left[\ln\left(\frac{8R}{r}\right) - 2\right] \tag{7.5}$$

μ_r is approximately 1 for air, while steel has a μ_r of approximately 2000 and iron has a μ_r of over 5000 depending on purity[1]. This is why most high value inductors have a center core made of iron.

Example 7.3.1: Calculate the inductance of a coil

A length of 24 gauge magnet wire is wrapped 100 times around an aluminum cylinder ($\mu_r \approx 1$) with a diameter of 1 cm. What is the inductance of the coil?

The inductance can be calculated by using equation (7.5). It is important to make sure all of the units are correct. When using $\mu_0 = 4\pi \times 10^{-7}\mathrm{H\,m^{-1}}$, the lengths all need to be measured in meters. 24 gauge wire has a diameter of 0.02 inches which converts to 508×10^{-6} m. We now have all of the variables we need:

- $N = 100$
- $\mu_r \approx 1$
- $R = 5 \times 10^{-3}$ m
- $r = 254 \times 10^{-6}$ m

By plugging in these values to the inductor coil equation (equation (7.5)), the inductance is found to be 192 µH. If the aluminum cylinder is replaced by a steel cylinder, μ_r increases to 2,000 and the inductance to 384 mH.

7.4 Current and voltage measurements of components

As we have seen, simultaneous measurements of both current and voltage through a component is useful when understanding how it works. For some components, plotting current vs. voltage is sufficient. However, for components like inductors and capacitors that have derivatives in their equations, time domain measurements better reveal their operation. A simple circuit for measuring current and voltage of a device in time is shown in figure 7.6.

For most oscilloscopes, a differential probe or instrumentation amplifier is required to properly measure the voltage across R_{SENSE}. This is because the ground clips are normally connected internally which eliminates the ability to use multiple reference grounds. If a differential probe is not available, the measurements can be made by using two probes in conjugation with the scope's math subtraction features.

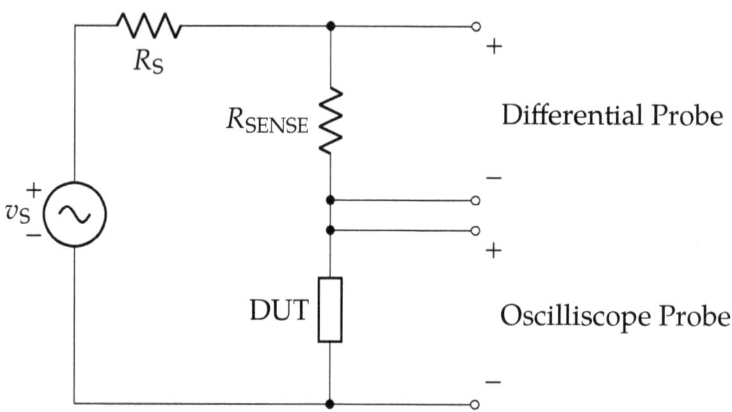

Figure 7.6: Circuit for simultaneous current and voltage measurements of an unknown device.

The current through the device under test (DUT) is determined from the voltage measurement over R_{SENSE} using Ohm's law. It is usually advantageous to use a resistor that is a power of 10 ($10\,\Omega$, $100\,\Omega$, $1\,\text{k}\Omega$, etc.) because the conversion from voltage to current can be done easily in your head.

Example 7.4.1: Calculating capacitance from time measurements

The circuit in figure 7.6 is built with an unknown capacitor and $R_{\text{SENSE}} = 100\,\Omega$. v_S is set to be a 10 kHz sine wave. The voltage across the DUT is measured by input 1, and the differential probe is connected to input 2. The measurement is shown in figure 7.7.

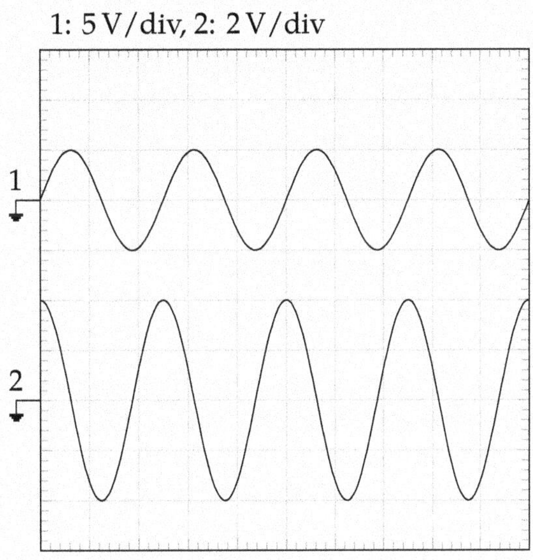

Figure 7.7: Oscilloscope measurement of circuit

The voltage across the capacitor is a 10 V peak to peak sine wave, and the voltage across the 100 Ω resistor is a 8 V peak to peak cosine wave. Using Ohm's law, we can determine the current through the resistor:

$$i_C = \frac{v_R}{100\,\Omega}$$

The current flowing through the capacitor is a 80 mA peak to peak cosine wave. Since we know that the capacitor must follow equation (7.2), we can compare it to the measurements and determine the capacitance.

$$\begin{aligned}
i_C(t) &= C\frac{dv_C(t)}{dt} \\
0.04\cos(2\pi 10\times 10^3 t) &= C\frac{d}{dt}5\sin(2\pi 10\times 10^3 t) \\
0.04\cos(2\pi 10\times 10^3 t) &= C\times 2\pi\times 50\times 10^3\cos(2\pi 10\times 10^3 t) \\
0.04 &= C\times 2\pi\times 50\times 10^3 \\
\frac{0.04}{2\pi\times 50\times 10^3} &= C \\
127.3\,\text{nF} &\approx C
\end{aligned}$$

7.5 LCR meters

LCR meters can be used to measure inductance, capacitance, and resistance of circuit elements. They are also capable of measuring the parasitic resistances of inductors and capacitors. A typical LCR meter has options for selecting the measurement frequency. Because real inductors and capacitors are not ideal, they do not have a constant inductance or capacitance over all frequencies. The measurement frequency should be set as close as possible to the frequency at which the circuit being built operates. This ensures that the measurement will predict the actual performance of the circuit.

Figure 7.8: A handheld LCR meter.

LCR meters are also useful for verifying the inductance of custom wound inductors and transformers. Custom winding of transformers is much more common than building custom capacitors.

7.6 Parallel and series capacitance and inductance

Much like we can simplify parallel and series networks of resistors, we can combine capacitors and inductors. Similarly, series connected inductors or capacitors form voltage dividers for time varying signals. Be careful to only combine capacitors with capacitors or inductors with inductors when simplifying circuits. Inductors and capacitors cannot be combined together using just the below rules![2]

[2] In future studies we will use the concept of *impedance* to apply series and parallel rules to inductors, capacitors, and resistors at the same time.

7.6.1 Combining inductors

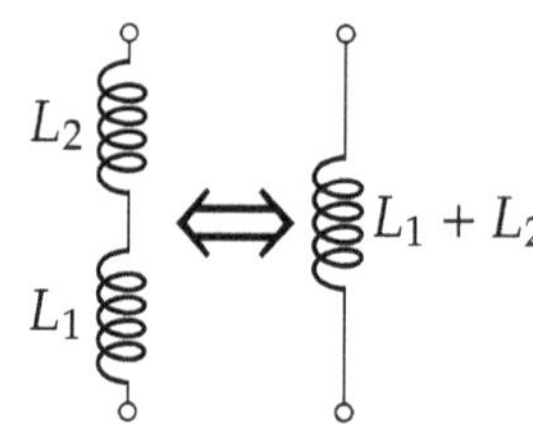

Figure 7.9: Series Inductors

The rules for inductors should look familiar because they are the same as resistors. The combined inductance of inductors in series is the sum of each inductor's inductance. For N inductors in series, each with inductance L_n, the total equivalent inductance is:

$$L_{\mathrm{eq,ser}} = \sum_{n=1}^{N} L_\mathrm{n}$$

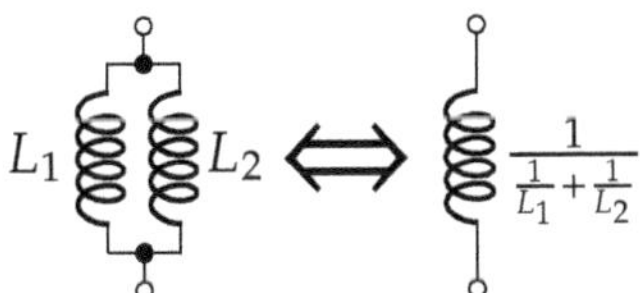

Figure 7.10: Parallel Inductors

Inductors in parallel follow the same rule as resistors in parallel. That is, the inverse of the equivalent inductance is equal to the sum of the inverse of each inductor's inductance.

$$\frac{1}{L_{\mathrm{eq,par}}} = \sum_{n=1}^{N} \frac{1}{L_{\mathrm{n}}}$$

7.7 Combining capacitors

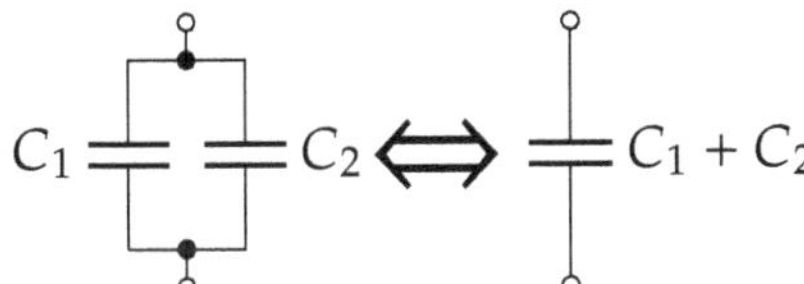

Figure 7.11: Parallel Capacitors

The rules for combining capacitors are complementary to the rules for inductors and resistors. That is, you add all of the individual capacitances together when capacitors are in parallel. N capacitors in parallel, each with capacitance C_{n}, would have an equivalent capacitance of:

$$C_{\mathrm{eq,par}} = \sum_{n=1}^{N} C_{\mathrm{n}}$$

An analogy is to treat capacitors as buckets of electrical charge. More buckets in parallel means more charge can be stored, so capacitors in parallel increase the total equivalent capacitance.

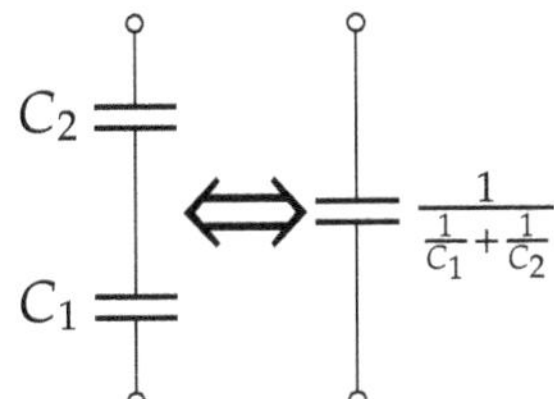

Figure 7.12: Series Capacitors

Capacitors in series follow a similar rule to parallel resistors. The inverse of the equivalent capacitance is the sum of the inverse of the capacitance of each capacitor. In this case, N capacitors in series each with capacitance C_n would have an equivalent capacitance that satisfies the following equation.

$$\frac{1}{C_{eq}} = \sum_{n=1}^{N} \frac{1}{C_n}$$

7.7.1 Voltage division

Inductors and capacitors will also act like voltage dividers when connected in series; however, there are some major drawbacks to both at low frequencies. For inductors, the voltage division equation is the same form as the voltage division equation for resistors:

$$v_{OUT} = \frac{L_1}{L_1 + L_2} v_{IN} \tag{7.6}$$

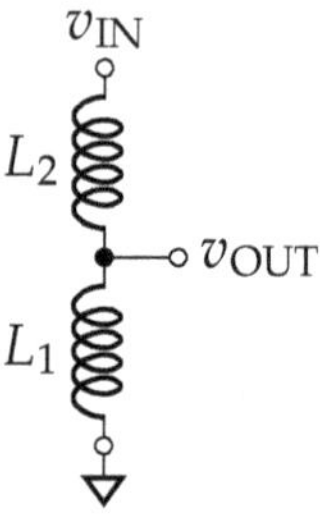

Figure 7.13: Inductive Divider

While equation (7.6) may appear straightforward, it is hiding a major issue. Recall that the characteristic equation for an inductor is $v_L = L\frac{di_L(t)}{dt}$. In order for the divider to develop enough voltage across each inductor to add up to v_{IN} and satisfy Kirchhoff's voltage law (KVL), there must either be a high frequency or high current. The large current at low frequencies make the inductive voltage divider hard to use in a practical application.

What happens when $\frac{di_L}{dt} = 0$? The inductor cannot have any voltage across it, so the current will reach a maximum only limited by the parasitic resistance of the inductors. This can make inductors catch fire!

Much like the parallel and series equations, the capacitive divider does not follow the same equation as the resistive or inductive divider. A two capacitor voltage divider will behave according to the following relationship:

$$v_{OUT} = \frac{C_2}{C_1 + C_2} v_{IN} \tag{7.7}$$

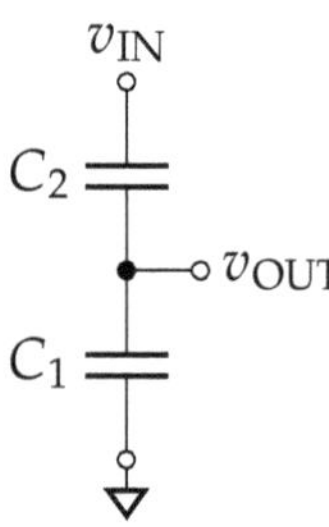

Figure 7.14: Capacitive Divider

Once again, equation (7.7) is hiding a potential flaw. From the capacitor characteristic equation, $i_C(t) = C\frac{dv_C(t)}{dt}$, we see that the current flow through a capacitor is directly linked to the amplitude and frequency of the voltage. If the frequency is low, the output of the divider cannot drive enough current to maintain the output voltage.

7.8 Prelab

Task 7.8.1: Prelab questions

1. How many turns would be needed to make a 1 mH inductor with the following specifications:
   - 18 AWG magnet wire (diameter of ≈ 1.02 mm)
   - 2 cm coil *diameter*
   - Air filled ($\mu_r = 1$)
2. The circuit in figure 7.15(b) is built using an unknown inductor. Determine the inductance using the measurements shown in figure 7.15(a). Input 1 is measuring the voltage across the inductor and input 2 is measuring the voltage across the resistor. The source is a 10 kHz sinusoidal voltage.

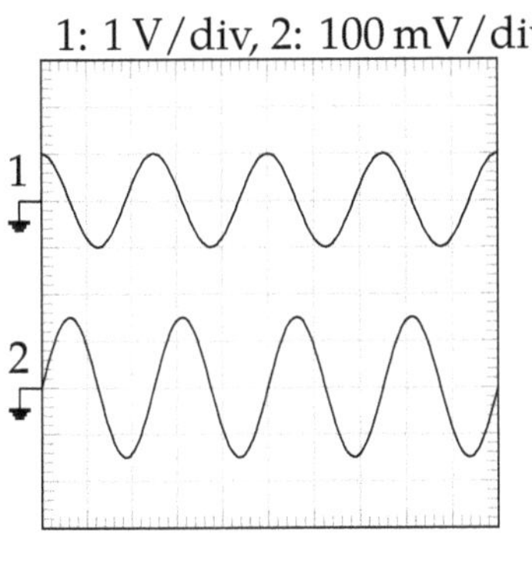

(a) Measurement results

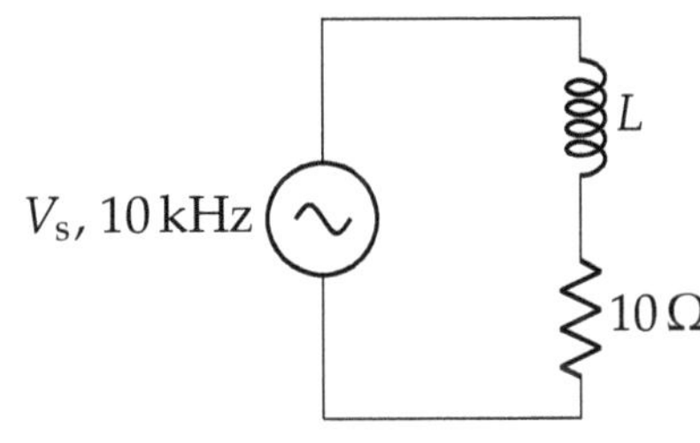

(b) Test circuit

Figure 7.15: Current and voltage measurement of an inductor.

7.9 Tasks

Task 7.9.1: Building resistors, capacitors, and inductors 

1. A piece of pencil lead (made of graphite) is a slightly resistive conductor.
   a) *Estimate* the resistance of the piece of pencil lead given that graphite has a resistivity on the order of $3 \times 10^{-5}\,\Omega\,\text{m}$.
   b) *Measure* the resistance of the piece of pencil lead using a DMM or LCR meter.
   c) *Calculate* the error between the estimated and measured resistance.
   d) *Calculate* the actual resistivity of the piece of graphite you used.
2. A simple parallel plate capacitor can be built using two sheets of a conductor.
   a) *Calculate* the capacitance between two metal plates separated by a single sheet of printer paper (about 0.1 mm thick). Use the size of metal plates provided by your instructor.[a]
   b) *Build* the parallel plate capacitor, and *measure* its capacitance using an LCR meter.
   c) *Calculate* the error between the predicted and measured capacitance.
3. An inductor can be built by tightly wrapping wire around a small metal rod (like a screwdriver).
   a) *Calculate* the inductance of a 20 turn air filled inductor with diameter equal to the rod you wrap it around.
   b) *Build* the inductor, and *measure* its inductance using an LCR meter.
   c) *Calculate* the μ_r for the rod used based on your inductance measurement.
   d) *Identify* the types of metal the rod could be made from based on the measured μ_r.
   e) Keep the hand wound inductor for later tasks.

[a] Be sure to look up the dielectric constant for paper!

Task 7.9.2: Current and voltage measurements of capacitors

1. *Build* the circuit in figure 7.6 with $R_{\text{SENSE}} = 1\,\text{k}\Omega$ and a 10 nF capacitor as the DUT.
2. *Apply* a 10 V peak to peak 100 kHz sine wave to the circuit, and *measure* the voltage across the DUT and the voltage across R_{SENSE}.
3. *Calculate* the capacitance of the capacitor using the current through R_{SENSE} and $I = C\frac{dV}{dt}$.
4. *Measure* the capacitance of the 10 nF capacitor using an LCR meter.

5. *Calculate* the error between this measurement and the LCR meter measurement.
6. *Repeat* using two 10 nF capacitors from your parts kit in parallel.
7. *Repeat* using two 10 nF capacitors from your parts kit in series.

Task 7.9.3: Current and voltage measurements of inductors

1. *Build* the circuit in figure 7.6 with $R_{\text{SENSE}} = 100\,\Omega$ and the hand wound inductor as the DUT.
2. *Apply* a 10 V peak to peak 100 kHz sine wave to the circuit, and *measure* the voltage across the DUT and the voltage across R_{SENSE}.
3. *Calculate* the inductance of the inductor using the current through R_{SENSE} and $V = L\frac{dI}{dt}$.
4. *Calculate* the error between this measurement and the LCR meter measurement.
5. *Repeat* using a 1 mH inductor with $R_{\text{SENSE}} = 1\,\text{k}\Omega$.

Task 7.9.4: Capacitor voltage divider

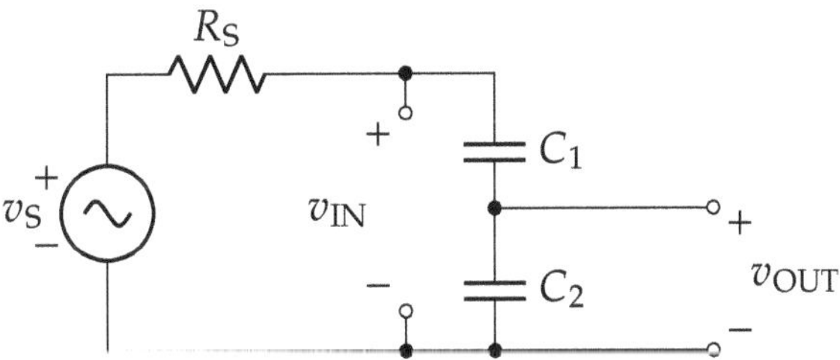

Figure 7.16: Capacitor voltage divider

1. *Construct* the circuit shown in figure 7.16 using $C_1 = 0.47\,\mu\text{F}$ and $C_2 = 0.1\,\mu\text{F}$.
2. *Set* the function generator to a 5 V peak to peak 10 kHz sine wave.
3. *Connect* one oscilloscope probe to the input and another one to the output.
4. *Capture* an oscilloscope screenshot showing both the input and output.
5. *Swap* C_1 and C_2, and *capture* a screenshot showing the new input and output.
6. *Sweep* the frequency of the function generator from 1 Hz to 1 kHz. How does the attenuation ratio change?
7. *Change* the function generator to DC mode and *sweep* the voltage from 0 V to 5 V. Does the circuit still work like a voltage divider?

7.10 References

[1] C. Balanis, *Engineering Electromagnetics*. Wiley, 1989, ISBN: 9780471621942.

EXPERIMENT 8

Operational Amplifiers: Basic Circuits

8.1 Application

The operational amplifier (op amp) is an integral tool for analog signal processing. It can be used in many different configurations to add, subtract, integrate, differentiate, and more. It is probably the most widely used type of integrated circuit (IC) in analog design. For reference, a small sample of historical uses include:

- analog computers that numerically solve complicated differential equations by realizing a circuit with the same equations. Generally, analog computers have been replaced by digital systems, but carefully designed op amp circuits can sometimes beat the accuracy or power consumption of the digital equivalent. This is incredibly important for the field of brain and biology inspired computing [1].
- analog control systems to adjust the position of machinery based on sensor inputs.
- signal conditioning to lower the effective resistance of a signal and/or scale and shift inputs to a level compatible with the next processing stage.

While digital circuitry has replaced many op amp applications, they are still essential at the analog to digital interface. In applications such as buffers for analog to digital and digital to analog converters, radios, and sensor signal conditioning, the op amp is still king.

Figure 8.1: A motor speed controller built mostly from op amps, resistors, and capacitors.

8.2 Operational amplifiers

At its core, the op amp is simply a two input amplifier with a large gain. The basic model for an op amp is shown in figure 8.2. For example, the LF356N op amp has a minimum open-loop gain(1) of 10^5. A device with this much gain may seem very useful; however, it is ultimately limited by the power supply it is connected to. In addition, the amount of gain provided by an op amp varies not only from chip to chip, but it also varies based on factors such as temperature. Instead of using the amplifier in "open-loop" mode, we can use feedback to design a circuit that behaves as desired.

(1) This is the gain of an op amp when no feedback is connected, often denoted A_v.

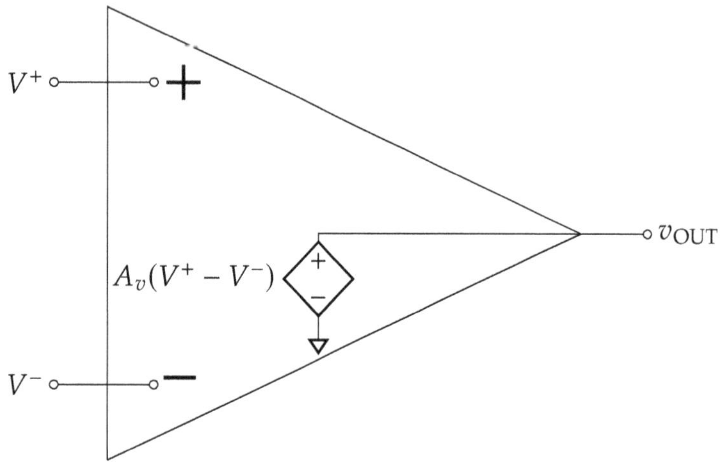

Figure 8.2: Model of an ideal op amp.

8.2.1 Open-loop performance

Op amps are not typically used as a traditional amplifier when no feedback is present. There are two major reasons: the gain is unpredictable, and the output distorts due to limitations of the power supply. Variations from chip to chip as well as temperature and frequency all affect the open-loop gain. The unpredictability of the open-loop gain makes design of consistent circuits difficult.

Power supply limitations also set a maximum voltage that can be used for an input that makes open-loop use impractical. For example, an op amp connected to a ±15 V power supply with gain of 10^5 will distort when the input signal $v^+ - v^-$ exceeds 0.15 mV(2). We very rarely deal with signals in the microvolts, so the open-loop configuration is not useful for a majority of signals.

(2) Actual op amps cannot reach the power supplies at the output, so this number would be even smaller in practice.

The open-loop op amp configuration does have one viable use: the comparator. The incredibly large gain of op amps drives the output to the positive output limit when $v^+ > v^-$ and the negative output limit when $v^- > v^+$.[3] That said, op amps should not be used as comparators because the outputs are not optimized for this type of use. A purpose built comparator should be used instead.

[3] Or, in this case when one is larger by about 0.15 mV.

8.2.2 The op amp buffer

An op amp follower circuit is shown in figure 8.3. The function of the circuit can be derived as follows

$$\begin{aligned} v_{\mathrm{OUT}} &= A_v(v^+ - v^-) \\ v_{\mathrm{OUT}} &= A_v(v_{\mathrm{IN}} - v_{\mathrm{OUT}}) \\ v_{\mathrm{OUT}} + A_v v_{\mathrm{OUT}} &= A_v v_{\mathrm{IN}} \\ v_{\mathrm{OUT}}(1 + A_v) &= A_v v_{\mathrm{IN}} \\ v_{\mathrm{OUT}} &= \frac{A_v}{A_v + 1} v_{\mathrm{IN}} \end{aligned} \tag{8.1}$$

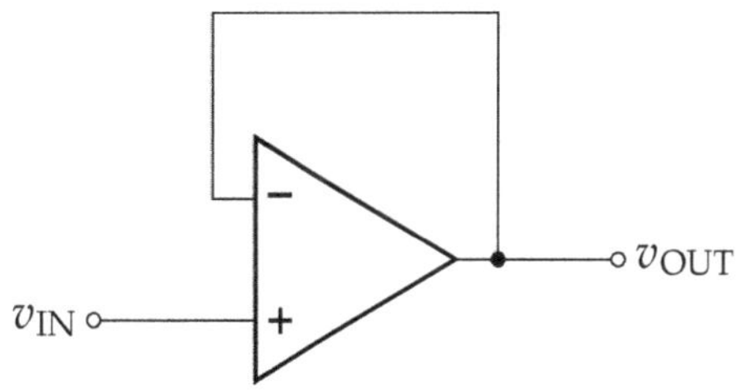

Figure 8.3: An op amp follower circuit.

Because A_v is large for an op amp, equation (8.1) simplifies to $v_{\mathrm{OUT}} \approx v_{\mathrm{IN}}$. Therefore, this circuit acts as a buffer for a signal because the output behaves like an ideal voltage source while the input draws no current[4].

[4] This is only completely true for an ideal op amp. Real op amps have a finite input and output resistance, so they aren't perfect buffers; however, a good op amp has very high input resistance (the LF356N has $10^{12}\,\Omega$) and quite low output resistance.

8.2.3 Virtual short circuit analysis

A commonly used method for analyzing circuits with large gain and negative feedback is called "virtual short circuit" analysis. Essentially, the negative feedback in op amp circuits works to minimize the difference in voltage between the positive and negative input terminals. Because of the feedback behavior, we say there is a virtual short circuit between the two inputs. This means that the terminals will both be at the same voltage and no current will flow between the two input nodes.

8.3 Common op amp circuits

There are several common op amp circuits that can be easily analyzed using the virtual short circuit method. Basic op amp circuits include the buffer, inverting amplifier, noninverting amplifier, and summing amplifier.

8.3.1 Inverting Amplifier

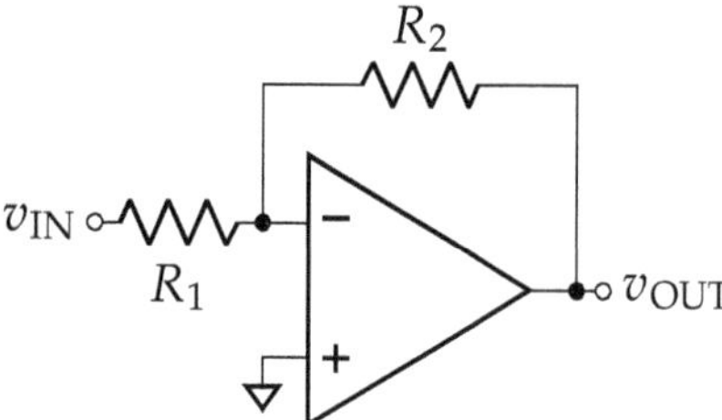

Figure 8.4: An inverting amplifier circuit

An inverting amplifier (figure 8.4) with a fixed gain (G) can be built using an op amp and two resistors as shown in figure 8.4. There is negative feedback present from v_{OUT} to the input, so virtual short circuit analysis can be applied. This causes v^- to be driven to 0 V. The circuit can then be solved by writing a single nodal equation at the v^- node.

$$\frac{v_{OUT} - v^-}{R_2} + \frac{v_{IN} - v^-}{R_1} = 0 \tag{8.2}$$

$$\frac{v_{OUT}}{R_2} + \frac{v_{IN}}{R_1} = 0$$

$$\frac{v_{OUT}}{R_2} = -\frac{v_{IN}}{R_1}$$

$$v_{OUT} = -\frac{R_2}{R_1} v_{IN} \tag{8.3}$$

From equation (8.3), we can see that the gain of the amplifier is $G = -\frac{R_2}{R_1}$.

Verification The virtual short circuit analysis can be confirmed by analyzing the circuit using the op amp dependent source model by inserting $v^- = v^+ - \frac{v_{OUT}}{A_v}$ into equation (8.2).

$$\frac{v_{OUT} + \frac{v_{OUT}}{A_v}}{R_2} + \frac{v_{IN} + \frac{v_{OUT}}{A_v}}{R_1} = 0$$

$$v_{OUT}\left(\frac{1 + \frac{1}{A_v}}{R_2} + \frac{\frac{1}{A_v}}{R_1}\right) = -\frac{v_{IN}}{R_1}$$

$$v_{OUT} = -\frac{R_2}{R_1 + \frac{R_1 + R_2}{A_v}} v_{IN}$$

In the limit as A_v gets large, the analysis using the model yields the same result as the virtual short circuit analysis.

Therefore, as long as $|A_v| \gg |G| = \frac{R_2}{R_1}$, then the inverting amplifier can be assumed to follow equation (8.3).

8.3.2 Non-inverting amplifier

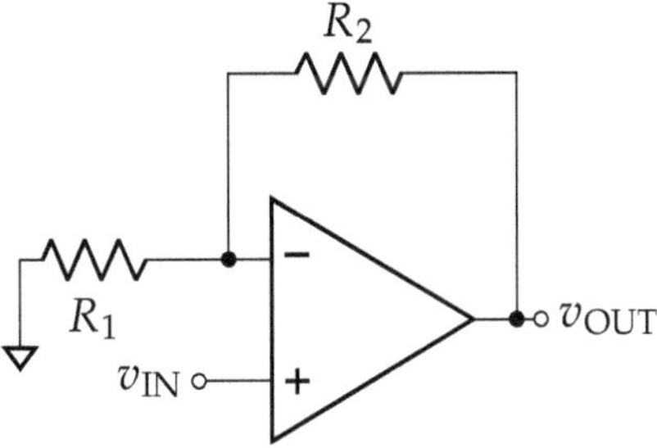

Figure 8.5: A non-inverting op amp circuit.

The non-inverting amplifier (figure 8.5) can also be solved by using a single nodal equation at v^-; however, the input is applied to v^+, so the virtual short circuit drives the node to $v^- = v_{IN}$.

$$
\begin{aligned}
\frac{v^- - 0}{R_1} &= \frac{v_{OUT} - v^-}{R_2} \\
\frac{v_{IN}}{R_1} &= \frac{v_{OUT} - v_{IN}}{R_2} \\
v_{OUT} &= \left(1 + \frac{R_2}{R_1}\right) v_{IN} \qquad (8.4)
\end{aligned}
$$

Note: this configuration cannot attenuate a signal since the minimum gain is one. If a noninverting gain of less than one is needed, then a resistor divider can be added at the input.

By inspection, the gain of the amplifier is $G = 1 + \frac{R_2}{R_1}$.

8.3.3 Summing amplifier

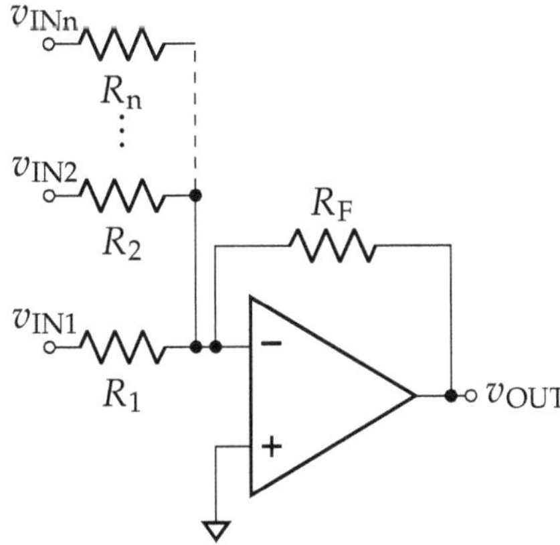

Figure 8.6: An N input summing amplifier circuit.

The inverting amplifier (figure 8.4) can be extended to combine and amplify multiple unique signals (figure 8.6). It can also amplify each signal separately before combining.

Analysis is done via superposition. For each input, treat the others as if they are connected to ground. Conveniently, the feedback drives v^- to ground, so no current flows through any resistors expect for where the remaining input is connected. This yields the result in equation (8.5), parameterized by i for the i-th input.

$$v_{\mathrm{OUT,i}} = -\frac{R_{\mathrm{F}}}{R_{\mathrm{i}}} v_{\mathrm{IN,i}} \tag{8.5}$$

The total result can be found by summing each superposition result from equation (8.5) together, yielding equation (8.6).

$$v_{\mathrm{OUT}} = \sum_{i=1}^{n} v_{\mathrm{OUT,i}} = -R_{\mathrm{F}} \sum_{i=1}^{n} \frac{v_{\mathrm{IN,i}}}{R_{\mathrm{i}}} \tag{8.6}$$

R_{F} controls the scaling of all inputs, while the input resistor R_{i} sets the gain (or attenuation) of the i-th input. This allows a designer to easily select independent gain values for each input to the op amp circuit.

8.4 Single-sided op amps

The popularity of mobile and battery power devices has made the single-sided op amp a commodity item in circuit design. These op amps are specifically designed to be operated with one voltage rail, rather than the two of a dual-supply device. Normally this voltage is positive, although, it is fine to use a negative supply as long as the ground potential is the more positive rail.

While single-sided op amps are convenient from a voltage supply prospective[(5)], they have additional issues that must be considered. Generally the solutions are relatively straight forward but require more additional components as compared to the equivalent dual-supply design. Two of the primary issues a designer will run into are:

(5) Creating a *quality* negative voltage from a battery is challenging.

1. **Only positive voltages**: An op amp can only swing the output between its supply voltages.[(6)] However, it is easy to imagine valid inputs that should result in an negative output and therefore not function correctly with a single-sided design. For example, an inverting amplifier with a positive input or a differential signal from a sensor system.

(6) In fact, real op amps are limited to something even less then that – called the *output voltage swing* or the *voltage range* in the datasheet. Some op amps limit the output to several volts away from the rail, while a few modern rail-to-rail output design can get within milivolts.

2. **Common mode voltage**: The common mode voltage of a differential signal(7) is the voltage "common" to both. Mathematically, this is the average of the two inputs and is stated for an op amp input as equation (8.7).

$$v_{\text{CM}} = \frac{v_+ + v_-}{2} \tag{8.7}$$

Circuit designs must respect an op amp's requirement on the input common mode voltage for proper operation.(8) For single-sided op amps, this requirement usually follows the form

$$0 \leq v_{\text{CM}} \leq V_{\text{DD}} - V_{\text{CM,hi}}$$

where $V_{\text{CM,hi}}$ is small with respect to V_{DD} but still large enough that one needs to consider it. Input common mode range is particularly important for low voltage designs.

(7) Like those applied across the two inputs of an op amp.

(8) Input common mode range is always in the op amp's datasheet.

8.4.1 Zero-signal reference

To more systematically address the issue of positive and negative signals in a positive (or negative) only system, we define a so called zero-signal reference. This is the voltage threshold in which anything less than it is considered "negative,"(9) and anything above is considered "positive". In other words, a two-sided signal, v, can be represented in is equivalent single-sided form, v_{ss}, by adding the zero-signal reference to it. That is,

$$v_{\text{ss}} = v + V_{\text{ref}}$$

where V_{ref} is the zero-signal reference voltage.

(9) Even though it is truly a positive voltage.

Similarly, the original signal can be recovered from its single-sided equivalent by taking

$$v = v_{\text{ss}} - V_{\text{ref}}$$

As we will see shortly, single-sided op amp circuits must carefully add direct, constant in time (DC) input biases such that the output has the desired zero-signal reference. However, we must also ensure that the input bias does not violate any input common mode voltage requirements in the process.

8.4.2 Input common mode voltage

As we have already seen, op amp circuits with negative feedback tend to force $v_+ = v_-$.[10] Therefore, going back to equation (8.7), it is clear that the common mode voltage is just the input itself.

$$v_{CM} = \frac{v_+ + v_-}{2} = \frac{2v_{\text{opamp,in}}}{2} = v_{\text{opamp,in}}$$

Where $v_{\text{opamp,in}} = v_+ = v_-$.

For the inverting op amp circuits in figure 8.4 and figure 8.6, the positive node v_+ is connected to the zero-signal reference, so the common mode input voltage can be controlled independently of the input signal; however, the buffer and noninverting amplifier do not offer the same luxury. The buffer and noninverting amplifier have the input signal connected to the positive op amp input, so the entire input signal must remain in the common mode input range at all times.

[10] For real systems, there is in fact a still a small difference between the nodes known as the *input offset voltage*. At typically several milivolts or smaller, it can often be safely ignored unless many op amps circuits are connected in series.

Dual-supply op amps

The ground node is the obvious choice for the zero-signal reference in a dual supply system.[11] Conveniently, ground is already available as a low impedance voltage to use in the circuit. It is no mistake that ground is bonded to one of the two op amps inputs in dual-supply designs. This results in $v_{\text{opamp,in}} = v_{\text{cm}} = 0\,\text{V}$ and satisfies nearly all input common mode requirements.

[11] Normally, dual-supply systems have equal and opposite voltage supplies, putting ground at exactly the middle of the supply voltage rails.

One major exception to the rule is the non-inverting amplifier, which has a $v_{CM} = v_{\text{in}}$, see figure 8.5. In this case, it is important to consider whether or not the input violates the input common mode requirements. However, much of the time the signal is zero-signal referenced to ground and the circuit has enough voltage gain that meeting the requirements is almost automatic.

Single-supply op amps

The lack of an obvious zero-signal reference makes things more difficult for single-sided designs. A common starting point is to maximize the symmetric swing by choosing the zero-reference as $V_{DD}/2$. However, this voltage is not normally already available as a low impedance node[12] and must be created as part of the design. In example 8.4.1, a voltage divider is used to create the required zero-signal reference and is heavily capacitively decoupled to provide a low impedance.

[12] A low impedance node is simply a node driven by a relatively low output impedance source (power supply, buffer, large capacitor, etc.).

Note: in general the resistor divider solution does not scale well because it must be repeated for each op amp in a circuit. For a circuit with many op amps, the reference voltage can be buffered or, if extreme precision is required, a dedicated voltage reference IC can be used.

Example 8.4.1: Single-sided op amp bias design

There are many ways to bias the input of an op amp to control the output DC value. One common method that works well with the inverting amplifier topology is shown in figure 8.7.

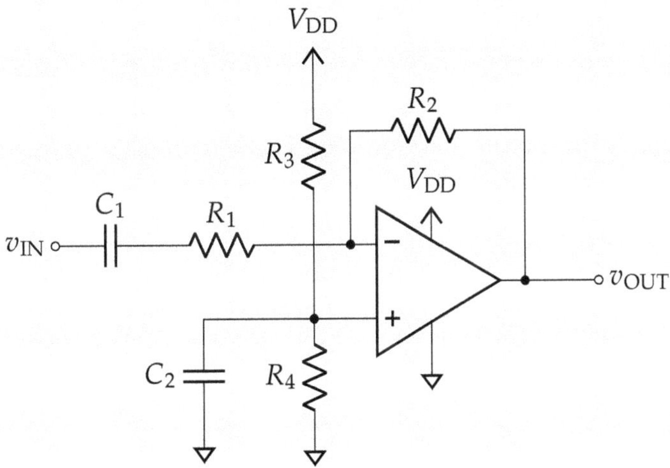

Figure 8.7: A single-sided inverting amplifier design with input biasing.

The easiest scheme to compute component values is to solve for how the circuit behaves with time varying signals and constant signals separately using an extension of superposition. We call the circuit model that holds for time varying signals the "AC equivalent circuit" and the version that holds for DC signals the "DC equivalent circuit."

Things are simple if we assume that the capacitors perfectly block DC (open circuit) and pass AC (short circuit) signals. The former is practically true and the later is true for AC signals of high enough frequency. The larger the capacitor, the lower the frequency needed to make the assumption. The resulting equivalent circuits are shown in figure 8.8.

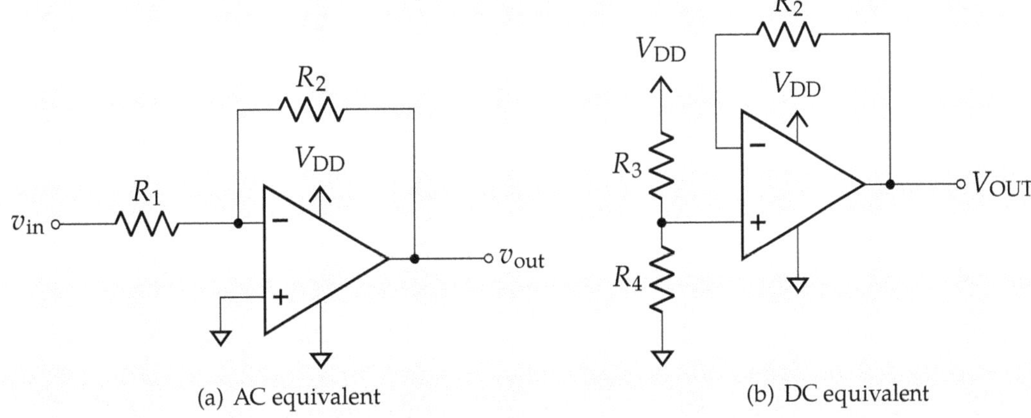

Figure 8.8: The AC and DC equivalent circuits of figure 8.7.

AC gain The AC equivalent circuit is found by replacing all of the capacitors with a short circuit. Notice that the equivalent circuit is the same as the standard inverting amplifier. Refer to section 8.3.2 for details on setting the AC gain.

DC gain The DC equivalent circuit is found by replacing all of the capacitors with open circuits. Because C_1 is replaced by an open circuit, we can also remove R_1 from the circuit. The output voltage in figure 8.8(b) is simply equal to the V_+ set by the resistor divider.

If it is not clear why R_2 doesn't effect the DC gain, then recall that no current flows into the negative input of the op amp. Therefore, there can be no DC current flowing through R_2. If no DC current flows through R_2, then there cannot be a DC voltage drop. Hence, $V_- = V_{\text{OUT}}$.

By the virtual short circuit assumption, we know that $V_- = V_+$. The voltage divider sets the positive input to $V_+ = \frac{R_4}{R_3+R_4}V_{\text{CC}}$. Finally, we arrive at the final form

$$V_{\text{OUT}} = \frac{R_4}{R_3 + R_4}V_{\text{CC}}$$

Using the same argument, the input common mode voltage is also equal to the DC voltage on the output. That is,

$$V_{\text{CM}} = \frac{R_4}{R_3 + R_4}V_{\text{CC}}$$

Selecting C₁ C_1 and R_1 form a high-pass filter on the input signal. A high-pass filter only allows fast signals through and blocks slower signals. In order to ensure that low frequency signals can pass through the amplifier, we must pick the cutoff point to be lower than the slowest signal of interest. For now, we will simply specify that the cutoff frequency is a tenth of the slowest input signal.

The cutoff frequency for an RC high-pass filter is $f_c = \frac{1}{2\pi RC}$. Therefore, if the slowest signal in the circuit is f_{min}, then we can solve for R_1C_1:

$$R_1C_1 > \frac{10}{2\pi f_{\text{min}}} \tag{8.8}$$

Selecting C₂ To ensure that fluctuations in the voltage rail do not propagate to the output and ruin the op amps high quality power supply noise rejection, we use C_2 to decouple the reference voltage. The larger the capacitance, the better the supply noise is rejected; however, if made too large, the capacitor's charge time will be so great that we would have to wait a long time for the circuit to arrive at its final bias.

A general guideline is to select the capacitor circuit's RC time constant such that the charge up time is reasonable in length while ensuring that signals at least 10 times slower than the slowest frequency of interest are sufficiently decoupled. The Thévenin equivalent resistance with respect to C_2 is equal to $R_3 \parallel R_4$. Therefore, the time constant is $\tau = C_2(R_3 \parallel R_4)$. A capacitor will be charged to 95% of the final voltage after 3τ,[a] so we can find an upper bound on C_2:

$$3\tau < t_{\text{charge,max}}$$
$$C_2 < \frac{t_{\text{charge,max}}}{3(R_3 \parallel R_4)}$$

We can use equation (8.8) to solve for a lower bound for C_2 as well:

$$C_2 > \frac{10}{2\pi f_{\min}(R_3 \parallel R_4)}$$

[a] Experiment 9 analyzes the step response behavior of a capacitor.

8.5 The comparator

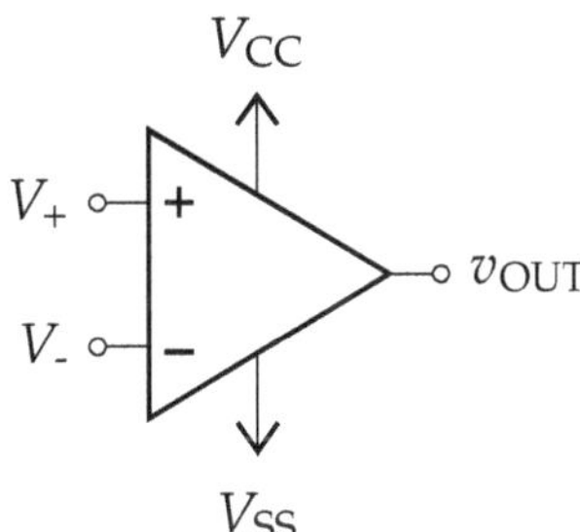

Figure 8.9: Basic comparator symbol.

It may bother you that the op amps and comparators share the same symbol. We find it slightly annoying too, since it causes students to grab op amps when they want a comparator. The reason is historical and is related to the fact that the internal circuitry is similar. Their primary difference is what the output stage is optimized for. An op amp is designed for outputs across the entire supply range while comparators are only designed to output the postive and negative rail voltages.

The comparator, with schematic symbol shown in figure 8.9, is a simple device that outputs a *digital* signal indicating which of the two *analog* inputs is larger. By digital we mean that the output is true, or binary one, when it is very near V_{CC} and false, or binary zero, when near V_{SS}. The generic comparator has the following input and output relationship:

$$V_{\text{out}} = \begin{cases} V_{CC} \text{ (Logic 1)}, & V_+ > V_- \\ V_{SS} \text{ (Logic 0)}, & V_+ < V_- \\ \text{unstable} & V_+ = V_- \end{cases} \tag{8.9}$$

In practice, the output may oscillate between V_{CC} and V_{SS} when $V_+ \approx V_-$ due to noise on either input rapidly changing which of the two signals is larger at any given moment in time.[13]

[13] Comparators with Schmitt triggers can be used to combat this problem.

8.5.1 The Open Collector/Drain Comparator

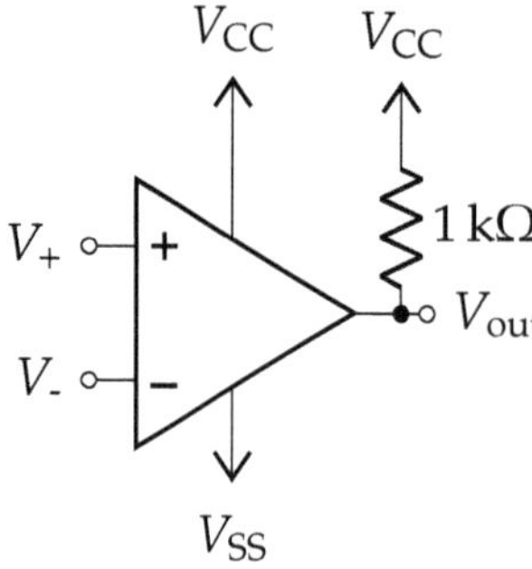

Figure 8.10: An open-collector comparator

Many comparators have what is called an "open-drain" or "open-collector" output.(14) What this means is that the output of the comparator is actually just an electronic switch connected to the V_{SS} pin of the comparator. There is no connection from v_{OUT} to V_{CC}, so the output can only sink current.

(14) It may not matter to your application, but collector vs. drain nomenclature indicates if bipolar junction transistors (BJTs) or metal-oxide-semiconductor field-effect transistors (MOSFETs) are used to implement the comparator.

In order to use this type of comparator, one should build the circuit in figure 8.10. The functionality of the circuit with the output resistor is the same as the standard comparator.

8.6 Prelab

Task 8.6.1: Prelab Questions

Please complete the following tasks

1. *Look up* the LF356N [2] and LM324N [3] datasheets.
   a) What is the range of supply voltages for each op amp?
   b) What is the input common mode requirement for each op amp?
2. *Calculate* the values for task 8.7.2 using the procedure outlined in example 8.4.1
   - *Select* R_1 and R_2 such that the AC gain is -10.
   - *Select* C_1 so that it passes signals of greater than 10 Hz without any significant impact on the overall gain using equation (8.8).
   - *Select* R_3 and R_4 so that the output DC bias is equal to $\frac{V_{DD}}{2} = 2.5\,\text{V}$.
   - *Select* C_2 such that the charge time is close to 1 s.

Task 8.6.2: Prelab task: the buffer amplifier circuit

You may use the Analog Discovery 2 to complete these tasks.

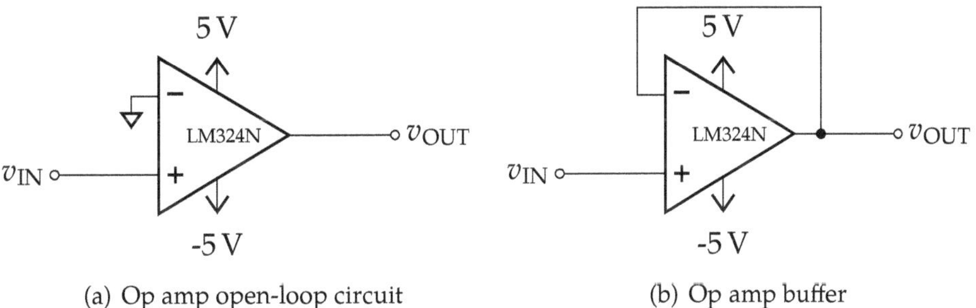

Figure 8.11: Open and closed loop op amp circuits.

1. *Build* the circuit in figure 8.11(a).
2. *Use* the function generator to generate $v_{IN} = 3\sin(2\pi 1000t)$.
3. *Capture* the input and output waveforms. Is the output an amplified version of the input? *Explain* why or why not.
4. *Build* the circuit in figure 8.11(b).
5. *Capture* the output waveform with the same input waveform. *Explain* why the output is different than the open-loop circuit.
6. *Increase* the input amplitude to $10\,\text{V}_{\text{p-p}}$. *Describe* what happens to v_{OUT}.

8.7 Tasks

Task 8.7.1: A non-inverting amplifier circuit

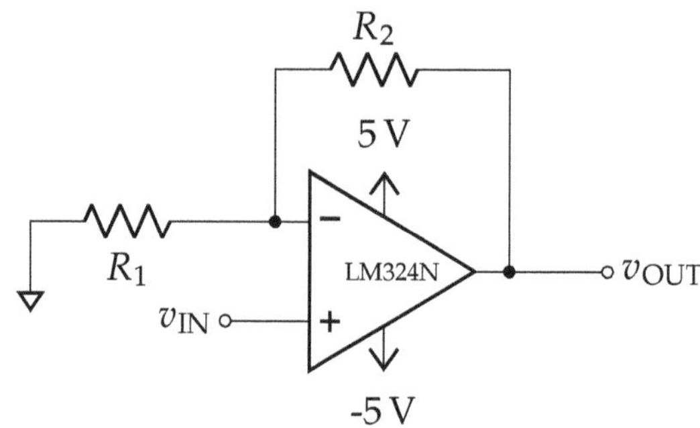

Figure 8.12: A non-inverting amplifier circuit.

1. *Calculate* appropriate values for R_1 and R_2 so that $v_{OUT} = 11v_{IN}$. Pick all resistances to be greater than $1\,\mathrm{k\Omega}$ to avoid loading the op amp.
2. *Construct* the circuit in figure 8.12 using the values you calculated for R_1 and R_2.
3. *Apply* a sine wave with an amplitude of 0.1 V at 1 kHz to v_{IN}.
4. *Capture* a printout showing v_{IN} and v_{OUT} on the same screen with peak to peak measurements for each channel.
5. *Calculate* the gain and the percent error.

Task 8.7.2: Usage of a single-sided op-amp

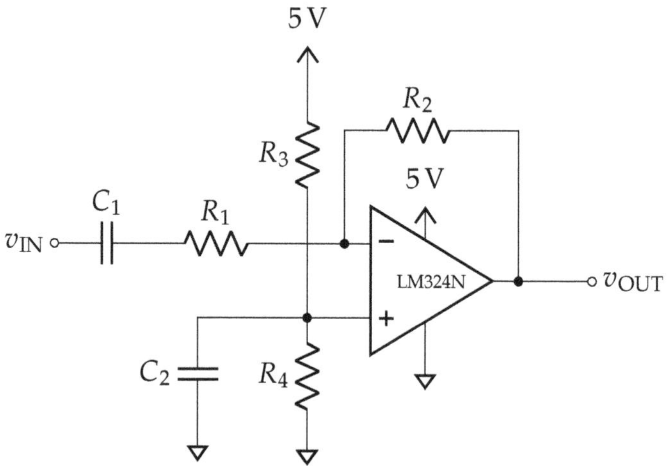

Figure 8.13: A single-sided amplifier circuit.

1. *Build* the circuit in figure 8.13 using the values calculated in the prelab.
2. *Apply* a reasonable v_{IN} based on the expected gain.
3. *Capture* an oscilloscope printout showing v_{IN} and v_{OUT} with measurements of the average and peak to peak values of the input and output.
4. *Compute* percent error for the AC and DC values measured.
5. *Adjust* the offset voltage of the source signal. Do you expect it to modify the amplifier's performance? *Capture* a set of printouts showing the new v_{IN} and v_{OUT}. Does the input DC offset have any effect?
6. *Verify* that a 10 Hz sinusoidal signal passes as expected. *Compute* the percent error for gain at 10 Hz.

Task 8.7.3: A comparator versus an op amp as a comparator

In this task you will compare the performance of an op amp as a comparator versus a device produced to be a comparator. While the schematic symbol is the same, real life performance is not.

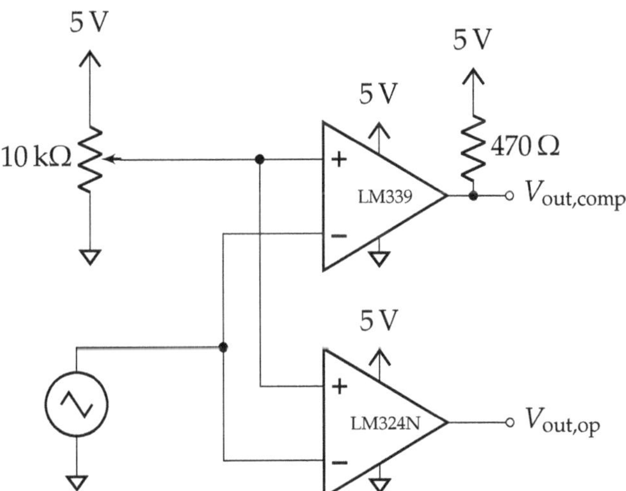

Figure 8.14: op amp versus comparator circuit.

1. *Build* the circuit in figure 8.14. *Adjust* the potentiometer to output approximately 2.5 V
2. *Set* the function generator to output a triangle wave that varies from 0 V to V_{CC} at 5 kHz.
3. *Capture* an oscilloscope screenshot showing both inputs and outputs at once. *Adjust* the display so that both the inputs are vertically aligned and both the outputs are vertically aligned and below the inputs. *Ensure* that the vertical scales are equal.

4. What is the difference between the LM324N as a comparator and the LM339 comparator? Does the op amp version perform as expected?

5. *Adjust* the potentiometer. What characteristic of the output waveform is changing? Why?

6. *Capture* a screenshot of an XY plot of v_{IN} and v_{OUT}. *Explain* the output.

7. *Set* the potentiometer output back to 2.5 V.

8. *Set* the oscilloscope to measure the rising edge count of the input and output and record the measurement.

9. Using the same base waveform as before, *add* noise to the FG output. *Set* the bandwidth to 1 MHz and the noise amplitude to 20% of the signal amplitude. Is the rising edge count as expected? Why or why not?

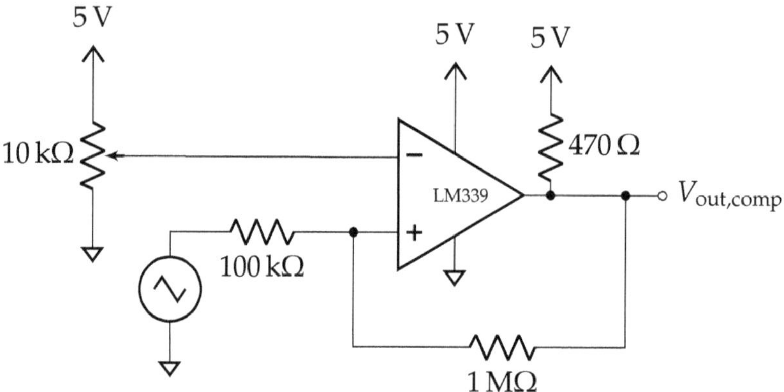

Figure 8.15: Adding hysteresis to a comparator circuit.

10. Using only the LM339 portion, *add* feedback resistors as shown in figure 8.15.

11. Using the same nosiy waveform, is the rising edge count as expected? Why or why not? In what way are the resistors effecting the reference voltage?

12. *Capture* an XY plot of v_{IN} and v_{OUT} without the additive noise.

13. *Adjust* the frequency to 0.1 Hz. *Describe* what happens when the input (X) signal is increasing and how it is different from when the input is decreasing.

    The effect seen is called hysteresis, and is a powerful way to limit the effect of noise.

8.8 References

[1] R. Sarpeshkar. (Jun. 2018). Analog supercomputers: From quantum atom to living body, TEDxDartmouth, [Online]. Available: https://www.youtube.com/watch?v=ZycidN_GYo0.

[2] *LFx5x JFET input operational amplifier*, LF356, SNOSBH0D, Texas Instruments Inc., May 2015. [Online]. Available: http://www.ti.com/lit/ds/symlink/lf356.pdf.

[3] *LMx24-N, LM2902-N low-power, quad-operational amplifiers*, LM324N, SNOSC16D, Texas Instruments Inc., Jan. 2015. [Online]. Available: http://www.ti.com/lit/ds/symlink/lm324-n.pdf.

EXPERIMENT 9

Step Response and Timing Circuits

9.1 Application

Step response is the reaction of a system after a sudden input, like a power supply turning on or a switch closing. Ideal resistive circuits will all respond instantly, but capacitors and inductors take time to react. This allows for designing circuits with new behaviors. The step response of a circuit can be used to design a class of timing circuits used for everything from audio synthesizers to motor controllers. The 555 timing circuit used in this lab is present in billions (yes, with a b) of devices worldwide.

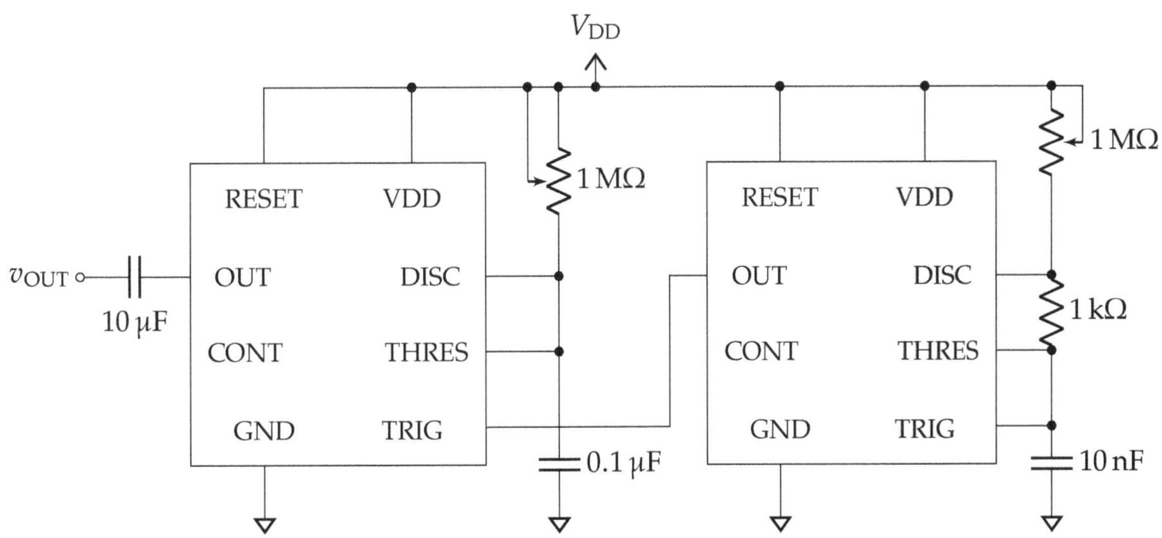

Figure 9.1: A simple stepped tone synthesizer built using two 555 timers. See it in action at https://youtu.be/Oi3dmSMpjsU.

9.2 Step response of capacitors and inductors

The step response of a circuit requires solving a first order differential equation of the following form:

$$\frac{\mathrm{d}x(t)}{\mathrm{d}t} + \frac{x(t)}{\tau} = \frac{y}{\tau}, \quad x\left(t_0^+\right) = x_0. \tag{9.1}$$

The solution to equation (9.1) must be a function that is a constant plus a scalar multiple of its first derivative. The only function that satisfies this condition is the exponential function $x(t) = ae^{t/b} + c$. The coeffecients a, b, and c are found by plugging the exponential function into equation (9.1). Altogether, the solution for $x(t)$ has the form

$$x(t) = x_0 e^{-\frac{(t-t_0)}{\tau}} + y\left(1 - e^{-\frac{(t-t_0)}{\tau}}\right). \tag{9.2}$$

x_0 is found using the initial conditions at $t = t_0^+$. The + sign indicates the time immediately after t_0. This is especially important when there are discontinuities in voltage (for inductors) or current (for capacitors), as we will see shortly.

Mathematically, $x\left(t_0^-\right)$ and $x\left(t_0^+\right)$ are shorthand for:

$$x\left(t_0^-\right) = \lim_{t \to t_0^-} x(t)$$

$$x\left(t_0^+\right) = \lim_{t \to t_0^+} x(t)$$

9.2.1 Capacitor step response

The step response for a capacitor is found by analyzing the Thévenin circuit in figure 9.2. The voltage source is a step function that turns on at time $t = t_0$.

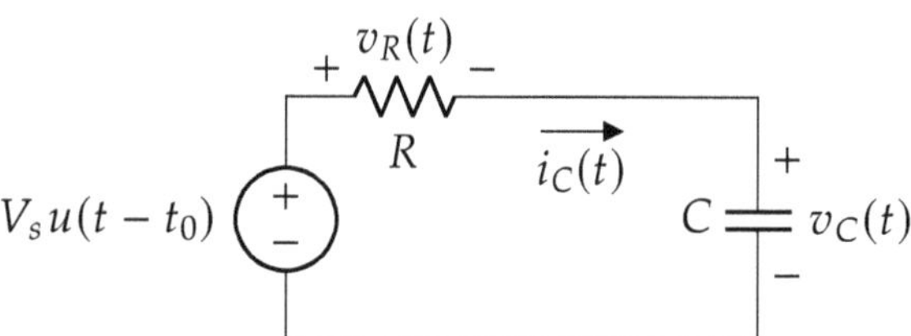

Figure 9.2: Resistor and capacitor step response circuit

A differential equation for the circuit can be obtained by applying Kirchhoff's voltage law (KVL).

$$\begin{aligned}
v_C(t) + v_R(t) &= V_s u(t - t_0) \\
v_C(t) + Ri_C(t) &= V_s u(t - t_0) \\
v_C(t) + RC\frac{\mathrm{d}v_C(t)}{\mathrm{d}t} &= V_s u(t - t_0) \\
\frac{\mathrm{d}v_C(t)}{\mathrm{d}t} + \frac{1}{RC}v_C(t) &= \frac{1}{RC}V_s u(t - t_0)
\end{aligned} \tag{9.3}$$

To deal with the step function in Equation (9.3), the problem is split into two. The regions and corresponding equations for $v_C(t)$ are

The unit step function $u(t)$ is defined as

$$u(t) = \begin{cases} 1 & t \geq 0 \\ 0 & t < 0 \end{cases}$$

$$\frac{dv_C(t)}{dt} + \frac{1}{RC}v_C(t) = 0 \qquad t < t_0 \qquad (9.4)$$

$$\frac{dv_C(t)}{dt} + \frac{1}{RC}v_C(t) = \frac{1}{RC}V_s \qquad t \geq t_0 \qquad (9.5)$$

Both equations (9.4) and (9.5) have the same form as the general differential equation in equation (9.1). Therefore, they both have solutions given by equation (9.2), where $\tau = RC$ and $y = 0$ or $y = V_s$, respectively.

If the capacitor is initially discharged[1] the initial condition is simply $V_c(-\infty) = 0$. Therefore, the solution to equation (9.4) is simply

$$v_C(t) = 0\,\text{V} \qquad t < t_0$$

Given that neither the source or the capacitor has a nonzero voltage, the current must also be zero,

$$i_C(t) = 0\,\text{A} \qquad t < t_0$$

The initial condition for equation (9.5) can be plucked directly from the previous result. Clearly at $t = t_0^-$, $v_C = 0\,\text{V}$. As we are working with a capacitor, $v_C(t_0^-) = v_C(t_0^+)$. In this case, the nonzero solution is

$$v_C(t) = V_s\left(1 - e^{-(t-t_0)/RC}\right) \qquad t \geq t_0 \qquad (9.6)$$

The capacitor's current is found by applying the I–V characteristic $i_C(t) = C\frac{dv_C(t)}{dt}$ or by looking at the current within the series resistance. That is

$$i_C(t) = C\frac{dv_C(t)}{dt} \qquad t \geq t_0$$

$$i_C(t) = \frac{V_s}{R}e^{-(t-t_0)/RC} \qquad t \geq t_0 \qquad (9.7)$$

or

$$i_C(t) = i_R(t) = \frac{V_s - v_C(t)}{R} \qquad t \geq t_0$$

$$i_C(t) = \frac{V_s}{R}e^{-(t-t_0)/RC} \qquad t \geq t_0$$

[1] If the circuit has be connected for a long time with the voltage source turned off, then the capacitor will be fully discharged by time $t = t_0$. This allows us to assume $v_C(t_0) = 0$ for many step response analyses.

On the other hand, if a capacitor has not been connected to a circuit for a long time, it may still contain charge! Always be sure to discharge large capacitors before storage or you may experience a *shocking* result the next time you use it.

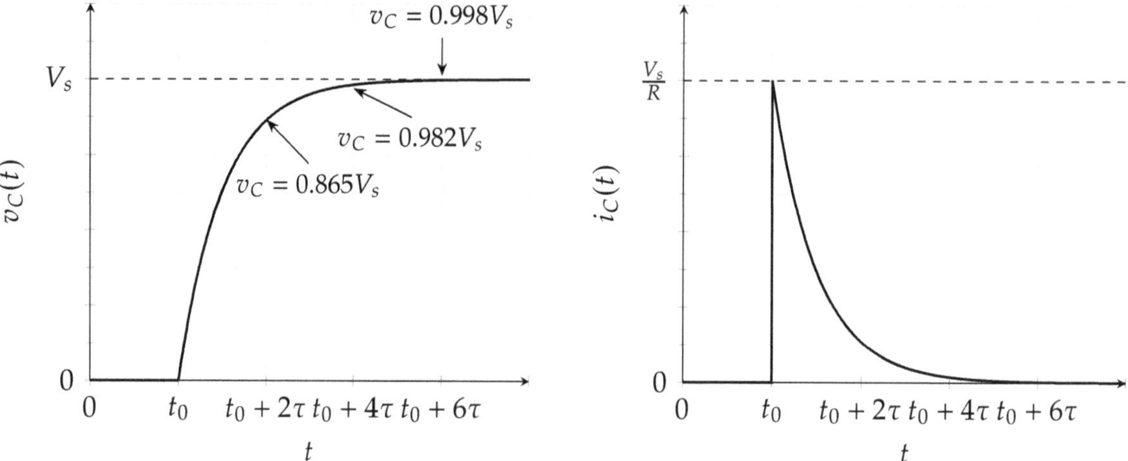

Figure 9.3: Step response of a capacitor

The plot of the step response is shown in figure 9.3. After 2τ, the capacitor has charged to 0.865 times the source voltage. After 6τ, the capacitor is within 1% of the source voltage.

Note that there is a discontinuity in the current waveform of the capacitor, so $i_C(t_0^-) \neq i_C(t_0^+)$. An important piece of intuition to form is that a capacitor allows for instantaneous changes in current but requires the voltage to remain continuous.(2)

(2) In actuality, the charge stored by a capacitor is continuous. As long as the capacitance remains constant, the voltage is continuous due to the relation $Q = CV$. Almost all capacitors have some change in capacitance based on applied voltage or temperature, but it can be safely ignored for these experiments.

9.2.2 Inductor step response

The procedure for finding the step response of an inductor is similar to the procedure for capacitor. It can be derived by analyzing the Thévenin circuit in figure 9.4.

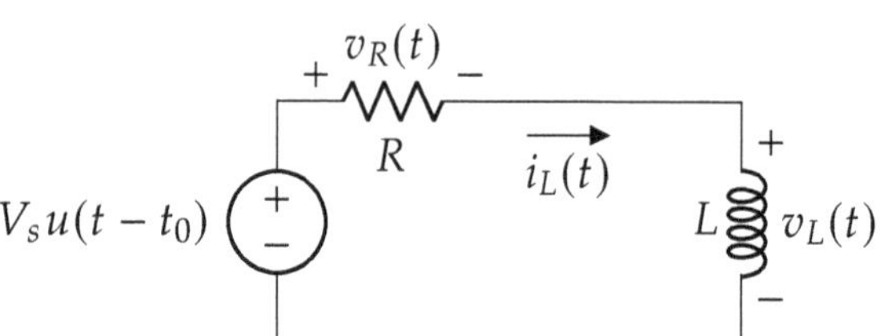

Figure 9.4: Resistor and inductor step response circuit

Equation (9.8) is found using KVL and substituting in $v_L(t) = L\frac{di_L(t)}{dt}$.

Unlike capacitors, it is near impossible to store energy in an inductor long term because there must be continuous current flow with no parasitic resistance in the loop. Unless you live somewhere that gets to −273 °C, you can safely assume the inductors in your kit will be discharged.

$$\frac{di_L(t)}{dt} + \frac{R}{L}i_L(t) = \frac{R}{L}\frac{V_s}{R}u(t - t_0) \tag{9.8}$$

The solution is found in the same way as it is for the capacitor. If we assume the circuit has been built for a long time prior to the step response, then the inductor will have the current

$$i_L(t) = 0\,\text{A} \qquad t < t_0$$

Similarly, the inductor voltage is also zero:

$$v_L(t) = 0\,\text{V} \qquad t < t_0$$

The initial current at the step is $i_L(t_0^-) = i_L(t_0^+) = 0\,\text{A}$, constraining the nonzero portion of the complete solution to

$$i_L(t) = \frac{V_s}{R}\left(1 - e^{-(t-t_0)\frac{R}{L}}\right) \qquad t \geq t_0 \qquad (9.9)$$

The inductor voltage can be found using $v_L = L\frac{di_L}{dt}$ or by subtracting the voltage across the resistor from the step voltage.

$$v_L(t) = V_s e^{-(t-t_0)\frac{R}{L}} \qquad t \geq t_0 \qquad (9.10)$$

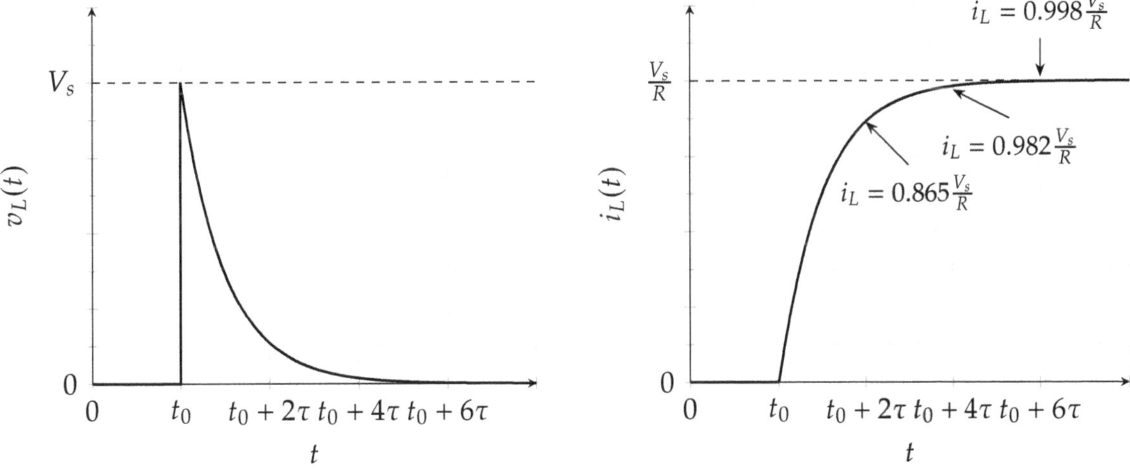

Figure 9.5: Step response of an inductor

The plot of the step response is shown in figure 9.5. For the inductor, the key property to understand is that the current remains continuous while allowing the voltage to change instantaneously.

9.2.3 General solution to step response problems

The previous step response solutions for inductor and capacitor circuits can be extended to any resistive network by first applying Thévenin circuit analysis. The general solution is given in equation (9.11).

$$x(t) = x\left(t_0^+\right)e^{-\frac{(t-t_0)}{\tau}} + x(\infty)\left(1 - e^{-\frac{(t-t_0)}{\tau}}\right) \tag{9.11}$$

$x(t)$ can be either current or voltage. $x\left(t_0^+\right)$ is the initial current or voltage at time t_0^+. t_0 is the time that the step function turns on or off. $x(\infty)$ is the final value of voltage or current at the limit as time goes to infinity. Finally, τ is either RC or L/R depending on whether the circuit contains a capacitor or inductor, respectively.

Capacitor and inductor boundary conditions

As a quick review, the voltage across capacitors and the current through inductors remain continuous. The following equations are true for any circuit where capacitance is constant[3]:

$$\begin{aligned} v_C\left(t_0^-\right) &= v_C\left(t_0^+\right) \\ i_C\left(t_0^-\right) &\neq i_C\left(t_0^+\right) \\ v_L\left(t_0^-\right) &\neq v_L\left(t_0^+\right) \\ i_L\left(t_0^-\right) &= i_L\left(t_0^+\right) \end{aligned}$$

[3] If capacitance changes, $Q_C\left(t_0^-\right) = Q_C\left(t_0^+\right)$ must be used with the relationship $Q = CV$, as $v_C\left(t_0^-\right)$ may not be equal to $v_C\left(t_0^+\right)$.

9.2.4 Calculating L or C from a step response curve

The step response of a capacitor or inductor circuit can be used to calculate the value of the inductor or capacitor if the resistance is known. This is done by estimating τ from a plot of voltage or current. There are two ways this can be done. The first is calculating an exponential regression of the data using a computational tool. The coeffecient of t in the exponent is equal to $1/\tau$. This method is the most accurate; however, it requires interfacing the measurement equipment with a computer.

The second method is performed by estimating τ by measuring how much time it takes to go from the initial voltage or current to 0.632 times the final voltage or current. If it is a capacitive circuit, then $\tau = RC$ so $C = \tau/R$. In an inductive circuit, $\tau = L/R$, so $L = \tau R$.

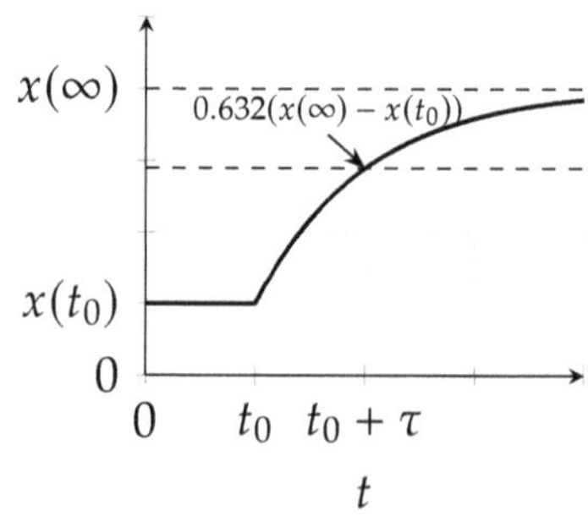

Figure 9.6: Finding τ from a step response measurement

Circuits for measuring τ

The step response can be measured using a function generator and oscilloscope assembled as shown in figure 9.7

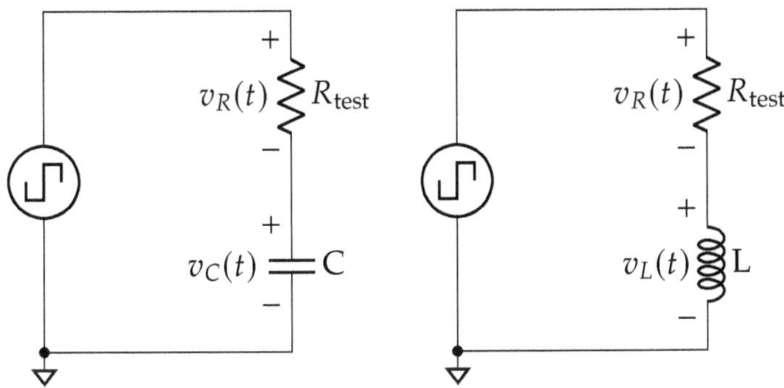

Figure 9.7: Test sets for step response

The period of the square wave should be much greater than 6τ in order to allow the device under test (DUT) to fully charge and discharge during a single period. Because τ is not known, the period should be set to be large and reduced if necessary.

τ can be calculated from the current or voltage response of the capacitor or inductor. The current can be computed by measuring $v_R(t)$ and dividing by the value of R. That is, $i_C(t) = \frac{v_R(t)}{R_{\text{test}}}$ or $i_L(t) = \frac{v_R(t)}{R_{\text{test}}}$.

9.3 Equivalent series resistance

Equivalent series resistance (ESR) is the estimation of energy dissipation within a device. Though we look at capacitors and inductors in schematics as if they are ideal, in practice there is an ESR for both. The ESR of inductors and capacitors comes from the fact that both of these are physical devices with physical properties. The properties can be estimated as a resistor in series with the ideal component, as shown in figure 9.8.

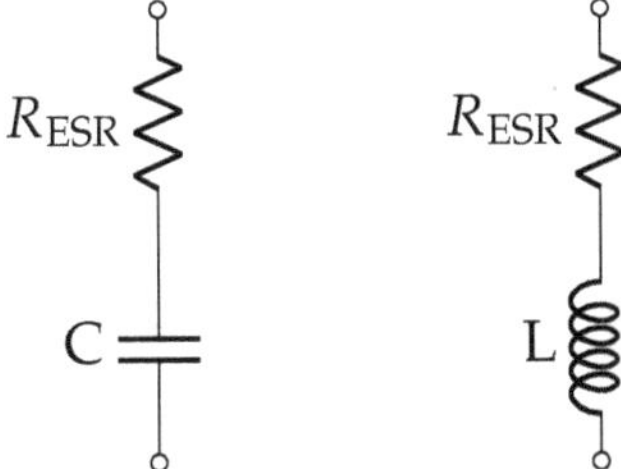

Figure 9.8: Inductor and capacitor ESR model

The ESR of inductors can be measured using a digital multimeter (DMM), as the inductor does not affect the measurement at DC. Measurement of capacitor ESR is more challenging. The simplest way is to use an LCR meter, but there are methods of measuring ESR using the step response.

9.4 Timing circuits

Timing circuits are used anywhere that something needs to happen for a specific amount of time or should repeat at a given frequency. A simple way to create timing circuits is to leverage the first order RC step response. The circuit for a simple RC oscillator is shown in figure 9.9.

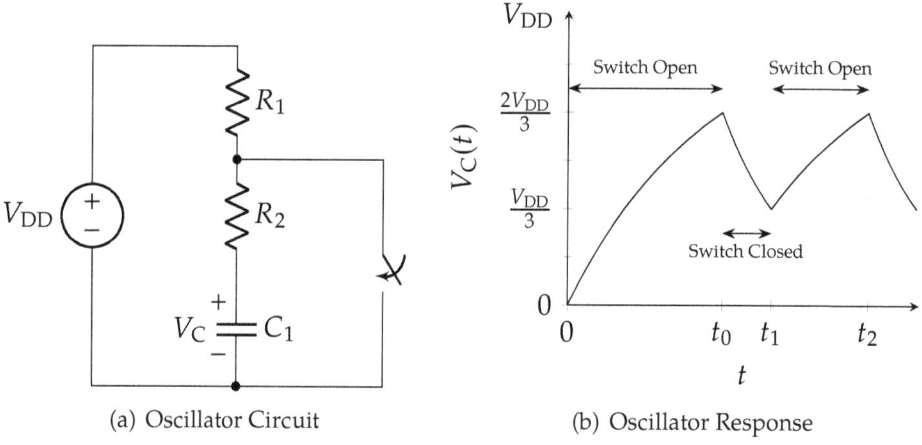

(a) Oscillator Circuit

(b) Oscillator Response

Figure 9.9: Simple Oscillator

In order for this circuit to oscillate, something needs to control when the switch opens and closes.(4) For now, we will simply analyze the circuit given that the switch closes when $V_C > \frac{2}{3}V_{DD}$ and opens when $V_C < \frac{1}{3}V_{DD}$. This results in the response shown by figure 9.9(b). In order to calculate the frequency that the circuit oscillates at, we independently analyze the two states of the circuit.

(4) One solution is part of the 555 timer described later in section 9.4.1.

At $t = t_0$, $V_C(t) = \frac{2}{3}V_{DD}$. We can use the initial condition to find the response of the circuit for $t_0 < t < t_1$. The switch is closed during this period, so the time constant is $\tau = R_2C_1$. The voltage supply and R_1 can be ignored as they are effectively disconnected from the capacitor circuit by the switch's short. The response of the undriven RC circuit is given in equation (9.12).

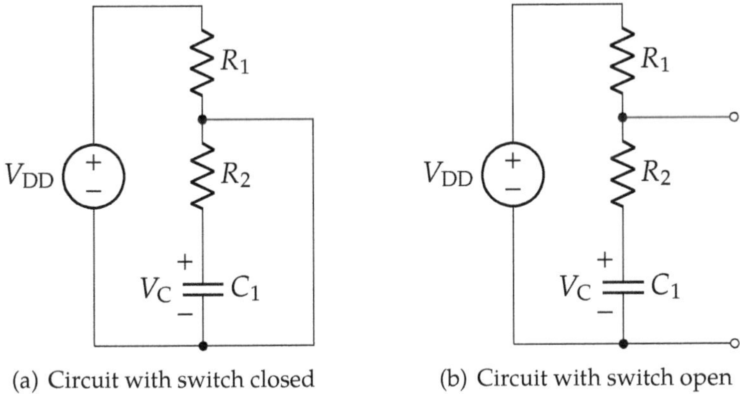

(a) Circuit with switch closed

(b) Circuit with switch open

Figure 9.10: Simple Oscillator States

$$
\begin{aligned}
v_C(t) &= V_C(t_0)e^{-\frac{t-t_0}{R_2C_1}} \\
v_C(t) &= \frac{2}{3}V_{DD}e^{-\frac{t-t_0}{R_2C_1}}
\end{aligned}
\tag{9.12}
$$

By setting $V_C(t_1) = \frac{1}{3}V_{DD}$, and then solving equation (9.12), we can find $t_1 - t_0$.

$$
\begin{aligned}
v_C(t_1) = \frac{1}{3}V_{DD} &= \frac{2}{3}V_{DD}e^{-\frac{t_1-t_0}{R_2C_1}} \\
-R_2C_2\ln\left(\frac{1}{2}\right) &= t_1 - t_0
\end{aligned}
\tag{9.13}
$$

Next, we need to find $t_2 - t_1$ using the circuit in figure 9.10(b). It is simply the RC circuit analyzed in section 9.2.1 with a resistance $R = R_1 + R_2$. Therefore, the voltage across the capacitor can be written as in equation (9.14) by plugging values into equation (9.11).

$$
\begin{aligned}
v_C(t) &= v_C(t_1)e^{-\frac{t-t_1}{(R_1+R_2)C_1}} + V_{DD}\left(1 - e^{-\frac{t-t_1}{(R_1+R_2)C_1}}\right) \\
v_C(t) &= V_{DD} - \frac{2}{3}V_{DD}e^{-\frac{t-t_1}{(R_1+R_2)C_1}}
\end{aligned}
\tag{9.14}
$$

The value of $t_2 - t_1$ is found by plugging $V_C(t_2) = \frac{2}{3}V_{DD}$ into equation (9.14).

$$
\begin{aligned}
V_C(t_2) = \frac{2}{3}V_{DD} &= V_{DD} - \frac{2}{3}V_{DD}e^{-\frac{t-t_1}{(R_1+R_2)C_1}} \\
-(R_1 + R_2)C_1\ln\left(\frac{1}{2}\right) &= t_2 - t_1
\end{aligned}
\tag{9.15}
$$

The total period of oscillation $t_2 - t_0$ can be found by adding equation (9.13) to equation (9.15)

Recall that $\ln\left(\frac{x}{y}\right) = \ln(x) - \ln(y)$

$$(t_2 - t_1) + (t_1 - t_0) = -(R_1 + R_2)C_1 \ln\left(\frac{1}{2}\right) - R_2C_1 \ln\left(\frac{1}{2}\right)$$

$$t_2 - t_0 = (R_1 + 2R_2)C_1 \ln(2) \tag{9.16}$$

The oscillation frequency of the circuit is just the inverse of the period given in equation (9.16).

$$f_{\text{osc}} = \frac{1}{(R_1 + 2R_2)C_1 \ln(2)} \tag{9.17}$$

9.4.1 The 555 timer

The 555 timer includes all of the necessary components to implement the oscillator circuit in figure 9.9. This simple timing application is part of the reason that the 555 timer is one of the most popular ICs ever built. In 2003 it was estimated that upwards of 1 billion units were sold per year [1]; that is over three 555 timers per United States citizen in the same year. The simple-to-use device can generate sinusoid, triangle, and square waves as well as implement pulse position modulation, one-shot timers, bounce free switches, frequency dividers, pulse width modulation (PWM) generators, and much more.

The timing chip has many variants, produced by several different manufactures. The original bipolar device was created by Signetics and marketed as the NE555 and SE555. Texas Instruments still sells the original design, along with the LM555 (bipolar), LMC555 (CMOS), and TLC555 (LinCMOS) variants. The list of variants goes on and on. Although the variants differ in specifications, they all operate with the same basic principals.

In general, the differences can be broadly summed up by saying that bipolar designs are able to deliver a larger output current than their CMOS equivalents; however, CMOS designs consume significantly less operating power and can operate at lower supply voltages. The maximum frequency of oscillation tends to be higher for CMOS, but is still limited to several MHz. Of course, these statements are only rough guidelines and one must verify any performance specification with the device's datasheet.

Understanding the operation of a 555 timer requires the understanding of two building blocks: The comparator and the set-reset latch (SR latch). We looked at comparators in experiment 8 and will consider the SR latch next.

The set-reset latch

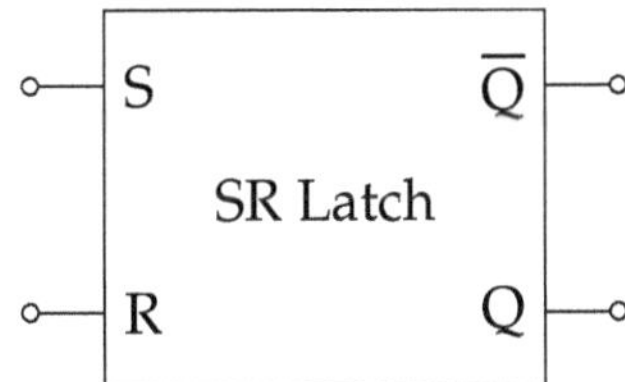

Figure 9.11: A basic set-reset latch (SR latch)

The set-reset latch (SR latch) symbol is shown in figure 9.11. In short, it is a two-gate digital circuit that is able to retain one bit of stored state. The latch's outputs Q and $\overline{Q}$ reflect this state and the logical compliment, respectively. The set (S) and reset (R) digital inputs can provoke a change by individually transitioning from low to high. Table 9.1 more clearly indicates these relationships.

S	R	Action	Q_{next}	$\overline{Q}_{next}$
0	0	hold	Q_{last}	$\overline{Q}_{last}$
0	1	reset	0	1
1	0	set	1	0
1	1	forbidden	X	X

Table 9.1: The state transitions matrix for an SR latch.

In words, and as seen in figure 9.12,

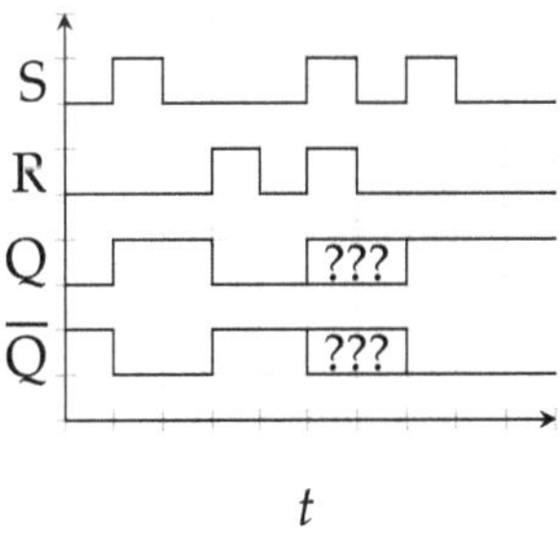

Figure 9.12: SR Latch Operation

- When both the set and reset inputs are low, the device maintains the current state and the outputs Q and $\overline{Q}$ remain constant.

- When the set input transitions to high and the reset remains low, the state is "set" to true and the outputs transition to reflect this. When the set pin returns low, the stored state remains a one.

- When the reset input transitions to high and the set remains low, the state is "reset" to false and the outputs transition to reflect this. When the reset pin returns low, the stored state remains a zero.

- If both "set" and "reset" are true, the output is indeterminate. This condition should be avoided and extra hardware is added to some latches to ensure that it never occurs.

9.4.2 General 555 timer operation

Most 555 timer applications leverage the charging time of an RC network to transition its output after a certain delay. The device's construction is almost as simple as its equivalent circuit shown in figure 9.13(5). Thankfully, it doesn't take much more effort to understand its operation. While the device may contain some unfamiliar digital components, their operation is simple. The primary analysis of the circuit's behavior can be done in the analog domain.

(5) In fact, you could easily breadboard your own 555 timer with three 5 kΩ resistors, two comparators, an SR latch, and a transistor. Try it if you really want to understand the 555!

To start our analysis of the operation, notice that the 555 timer has three 5 kΩ resistors(6) in series from V_{DD} to ground. This creates two voltages, $2/3V_{DD}$ and $1/3V_{DD}$, that are used as references for comparison to the trigger (TRIG) and threshold (THRES) inputs, respectively. The outputs of the comparators drive the S and R inputs of the internal SR latch and ultimately control the output voltage. Finally, an important detail to most of the 555 timer circuits is the discharge (DISC) pin. This pin is either in a high impedance state or connected directly to ground depending on the position of the switch controlled by $\overline{Q}$. That is, when $\overline{Q}$ is high (Q is low), DISC is connected to ground; otherwise, it is high impedance (essentially an open circuit).

There is a commonly held belief that the 555 timer is named for the three series 5 kΩ resistors; however, the 555 inventor, Hans Camenzind, said that the name was arbitrarily selected by the Signetics marketing manager [1].

(6) It is likely that modern CMOS variants use much larger resistor values; however, the net result is the same.

The control (CONT) pin is used to decouple the internal

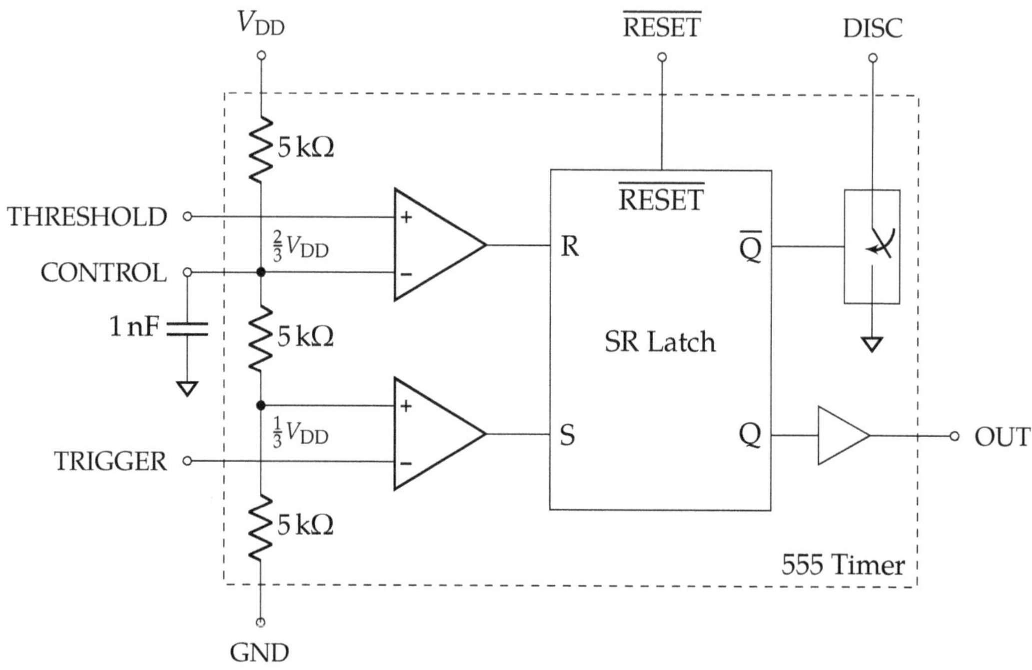

Figure 9.13: An equivalent circuit of the 555 timer IC

reference voltages. Most 555 timer datasheets call for a 1 nF capacitor to ground (GND), which should always be installed for proper operation[7]. In some 555 circuits, the control pin is used to change the threshold voltage of the comparators.

[7] In addition, decoupling capacitors should be added from V_{DD} to ground. You should always include decoupling capacitors for every IC in your circuit and ensure that the capacitors are as physically close the ICs pins as possible.

The active-low external reset ($\overline{\text{RESET}}$) pin resets the SR latch regardless of the current TRIG and THRES input voltages. While the external reset pin can be very useful, we will not need it for our purposes in this experiment. We will simply connect the pin to V_{DD}.

State transitions

Looking back to the SR latch, we know that the output (OUT) can only transition if either the S or R input is pulsed. Therefore, given the internal wiring of the 555 timer, the output can only change if one of two things occur[8] (see figure 9.14):

[8] Assuming the $\overline{\text{RESET}}$ pin is tied to V_{DD}

- The TRIGGER pin voltage decreases to below one-third of V_{DD}. In this case, the comparator will transition from driving the S pin low to driving it high, thereby pulsing the S input and causing the SR latch to set. The output goes high, and the switch opens.
- The THRESHOLD pin voltage rises above two-thirds of V_{DD}. In this case, the comparator transitions from driving the R pin low to driving it high, thereby pulsing the R input and causing the SR latch to reset. The output goes low, and the switch closes.

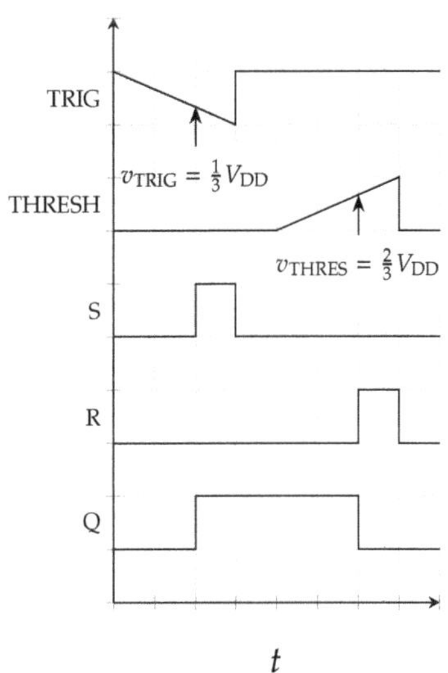

Figure 9.14: 555 Operation

The other possible changes in the comparator outputs will only result in a falling edge on the S or R inputs, which will not affect the stored state of the SR latch. Recall that an input of S=0 and R=0 results in the next Q being equal to the current one. However, the designer must still be careful to ensure that the external network driving the TRIG and THRES pins will not cause internal S and R pins to simultaneously be driven high. Normally this is not much of a concern because most 555 timer circuits supply the TRIG and THRES pins through the same RC network, ensuring only one of the two latch inputs can be activated at a time.

555 timer circuits

Typical 555 timer circuits can be generally broken down into one of three operating modes: astable, monostable, and bistable. All three modes are widely utilized for a variety of applications.

Astable mode In this mode, the 555 timer operates like an oscillator, see figure 9.15. Typical applications include LED and lamp flashers, clock signal generation, buzzer alarm generators, and more.

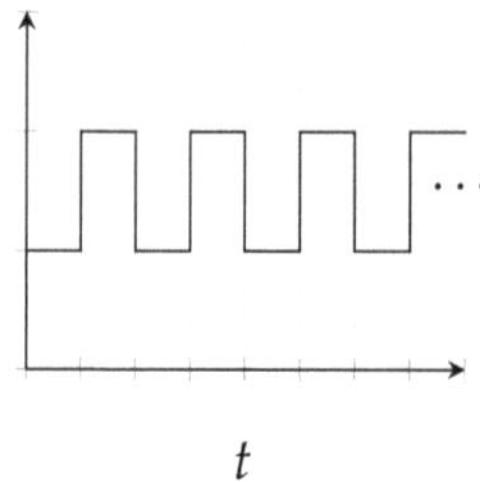

Figure 9.15: Astable output

Monostable mode In this mode, the 555 timer generates a single "one-shot" pulse that lasts a fixed amount of time, see figure 9.16. It can be used for many applications including pulse detection, switch debouncing, and frequency division.

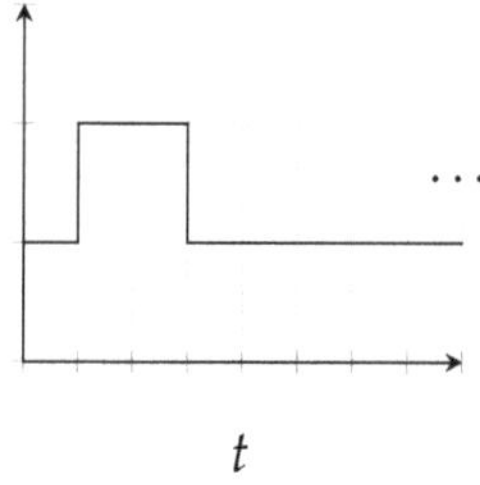

Figure 9.16: Monostable output

Bistable mode In this mode, the 555 timer operate as a flip-flop. It can also be used for switch debouncing.

It is difficult to appreciate the 555 timer without looking at some of its applications. You can build and simulate some 555 timer circuits in [2] to get a sense of different operation modes. In addition, the remainder of the section will focus on an illustrative example that should drive home the basic working theory and provide you the necessary tools to understand most other 555 timer circuits. We will explore a second topology in the tasks.

Duty cycle

Duty cycle is a measurement for square waves that determines the percentage of time that a signal is on compared to off. Typically, it is measured as the time on divided by the whole period; however, some data sheets may use time off divided by the period. For the square wave in figure 9.17, the duty cycle is

$$D = \frac{t_{\text{on}}}{T} = \frac{t_{\text{on}}}{t_{\text{on}} + t_{\text{off}}} = \frac{1}{3}$$

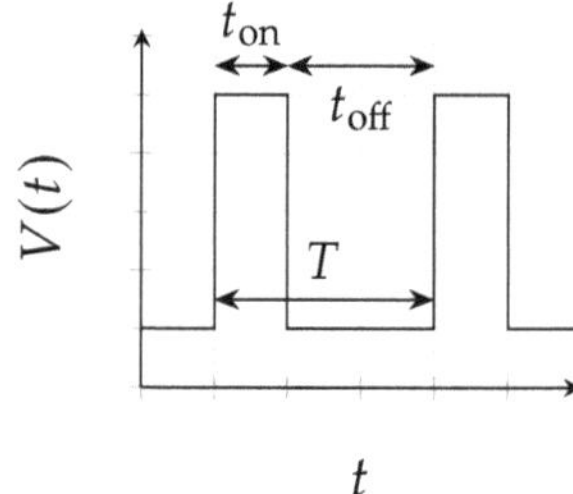

Figure 9.17: Square wave

Example 9.4.1: 555 timer operation in astable mode

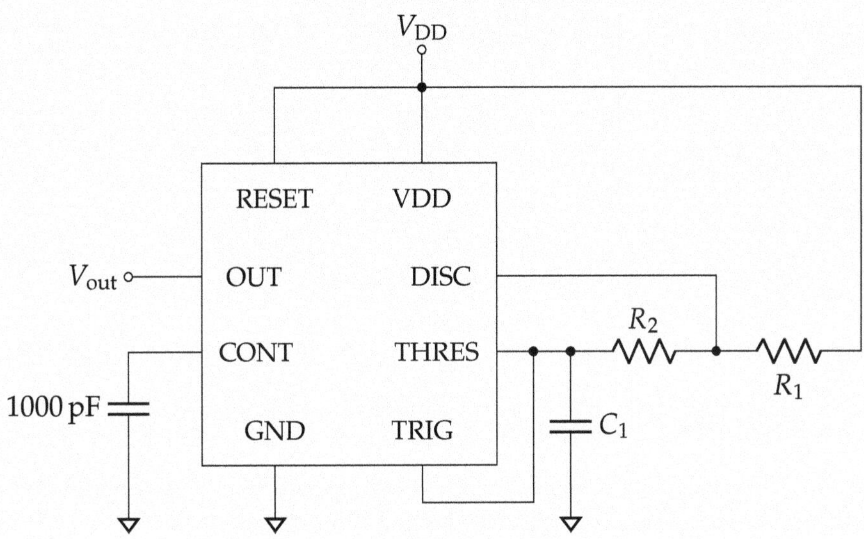

Figure 9.18: An astable mode 555 timer circuit.

Figure 9.18 shows the schematic for astable mode operation of a 555 timer. In this mode the external capacitor C_1 charges and discharges between $\frac{1}{3}V_{DD}$ and $\frac{2}{3}V_{DD}$. The equivalent circuit is equal to the previously analyzed circuit in figure 9.9(a).

When V_{DD} is first applied to the circuit, the wire to the DISC (discharge) pin is open and the capacitor charges through R_1 and R_2. The equivalent circuit is shown in figure 9.10(b) on page 115. We can find the time it takes to charge from $\frac{1}{3}V_{DD}$ to $\frac{2}{3}V_{DD}$ by solving the following for t_1:

$$v_C(t_1) = \frac{2}{3}V_{DD}$$
$$V_{DD} - \frac{2}{3}V_{DD} \cdot e^{\frac{-t_1}{RC}} = \frac{2}{3}V_{DD}$$

where

$$R = R_1 + R_2$$
$$C = C_1$$

After some manipulations, we get:

$$t_1 = \ln 2 \cdot RC = \ln 2 \cdot (R_1 + R_2) \cdot C_1 \approx 0.693 \cdot (R_1 + R_2) \cdot C_1$$

When the threshold voltage reaches $\frac{2}{3}V_{DD}$ or greater, the discharge pin is short circuited to ground, and the capacitor begins to discharge.

The new equivalent circuit is the same as in figure 9.10(a) on page 115. The time it takes to discharge can be found by solving the following for t_2:

$$\frac{2}{3}V_{DD} \cdot e^{\frac{-t_2}{RC}} = \frac{1}{3}V_{DD}$$

where

$$R = R_2$$
$$C = C_1$$

We then get:

$$t_2 = -\ln\frac{1}{2} \cdot RC \approx 0.693 \cdot R_2 \cdot C_1$$

The total time needed to charge and discharge the capacitor can be found by summing up t_1 and t_2:

$$T = 0.693 \cdot (R_1 + 2R_2) \cdot C_1$$

This time is how long it takes to go through one period of the oscillating waveform, so the frequency of the output waveform can be calculated as:

$$f = \frac{1}{T} = \frac{1.44}{(R_1 + 2R_2) \cdot C_1}$$

In addition, since the external capacitor charges through R_1 and R_2 and discharges through only R_2, the duty cycle may be set by the ratio of these two resistors. Specifically, the duty cycle can be described as:

$$D = \frac{t_1}{t_1 + t_2} = \frac{R_1 + R_2}{R_1 + 2R_2}$$

From this the value of R_1, R_2, and C_1 can be carefully adjusted to generate waveforms with a duty cycle at a certain frequency. For instance, if the square wave is to have 66% duty cycle at a frequency of 1 kHz, solve:

$$D = \frac{R_1 + R_2}{R_1 + 2R_2} = \frac{2}{3}$$

which implies,

$$R_1 = R_2$$

For this example, we pick the standard resistor $R_1 = R_2 = 4.7\,\text{k}\Omega$.

Then, for choosing the capacitor, we can solve:

$$f = \frac{1.44}{(R_1 + 2R_2) \cdot C_1} = \frac{1.44}{(4.7\,\text{k}\Omega + 9.4\,\text{k}\Omega) \cdot C_1} = 1\,\text{kHz}$$

The closest standard capacitor is 0.1 μF. This yields an oscillation frequency of 1.02 kHz. While this is not exactly what we wanted, the tolerances of the resistors and capacitors will likely cause more error than the slight design error. If a more precise frequency is needed, adjustable resistors and capacitors can be used; however, the 555 timer is not the correct tool for very high precision oscillators. For those cases, some sort of crystal oscillator circuit would be better.

9.5 Prelab

Task 9.5.1: Prelab questions

1. *Draw* the complete schematic for the circuit shown in figure 9.21 by replacing the 555 timer with the model of the 555 timer in figure 9.13.
2. *Calculate* the time it takes for $V_C(t)$ to charge to the 555 timer threshold voltage given that the capacitor is initially discharged. Your answer should be in terms of R_1 and C_1.
3. *Design* the astable 555 circuit in task 9.6.3 to have a frequency of 1 Hz and duty cycle of 60%.

Task 9.5.2: Prelab Task: Capacitor Step Response

You can use the Analog Discovery 2 to complete these tasks.

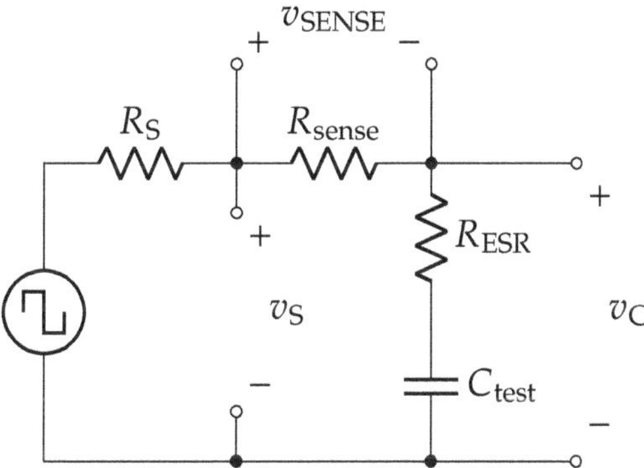

Figure 9.19: Simple RC circuit for capacitance measurement.

For this task, use the following values:

- $C_{test} = 2.2\,\mu F$
- $R_{sense} = 100\,\Omega$

1. *Measure* the capacitance and R_{ESR} of C_{test} using the LCR meter.
2. *Measure* the sense resistor R_{sense} using the LCR meter.
3. *Construct* the circuit in figure 9.19 with the given values.
4. *Use* the function generator to generate a 0 V to 2 V square wave at a reasonable frequency and *apply* it to the circuit.
5. *Measure* v_S and v_C using the oscilloscope then *adjust* the frequency so that the capacitor fully charges every period. *Set* the input square wave to be the trigger source.

6. *Setup* the oscilloscope math mode to calculate v_{SENSE} from v_S and v_C.
7. *Use* cursors to estimate τ by finding how long it takes the voltage to charge to 0.632 times the final value minus the initial value.
8. *Capture* an oscilloscope screenshot showing v_C, v_{SENSE}, and the measurement cursors.
9. *Estimate* C_{test} using the measured τ. Use $R_{\text{th}} = R_{\text{sense}} + R_S$.[a]
10. *Compute* the percent error of the calculated and measured capacitance.
11. *Export* the oscilloscope data to your computer.
12. *Use* Python's NumPy, MATLAB®, or other plotting tool to plot the measured voltage data and the ideal step response on the same plot. Repeat for the measured and ideal current response. Does the measurement match with theory? *Explain* any difference.

[a] $R_S \approx 0\,\Omega$ for the Analog Discovery 2 when the peak output current is less than 20 mA. The output resistance of the Analog Discovery 2 increases rapidly when loaded above 20 mA. The benchtop function generator will have $R_S = 50\,\Omega$ regardless of load.

9.6 Tasks

Task 9.6.1: Inductor step response

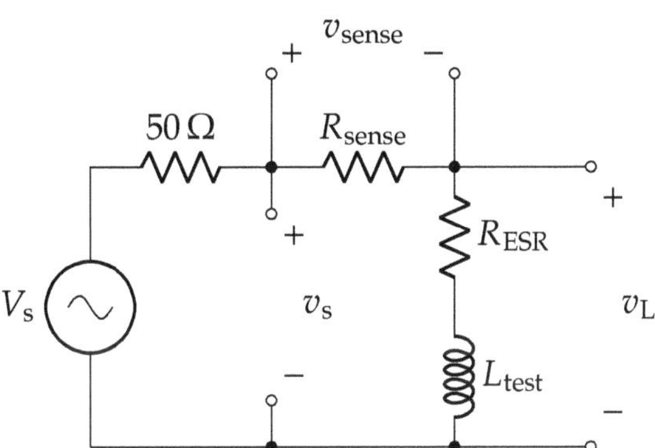

Figure 9.20: Simple RL circuit for inductance measurement.

For this task, use the following values:

- $L_{\text{test}} = 1\,\text{mH}$
- $R_{\text{sense}} = 100\,\Omega$

1. *Measure* the inductance and R_{ESR} of L_{test} using the LCR meter.
2. *Measure* the sense resistor R_{sense} using the LCR meter.
3. *Construct* the circuit in figure 9.20 with the given values.
4. *Use* the function generator to generate a $-1\,\text{V}$ to $1\,\text{V}$ square wave at a reasonable frequency and *apply* it to the circuit.
5. *Measure* v_{S} and v_{C} using the oscilloscope, and then *adjust* the frequency so that the capacitor fully charges every period. *Use* the input square wave as the trigger source.
6. *Set* the oscilloscope math mode to calculate v_{SENSE} from v_{S} and v_{C}.
7. *Use* cursors to estimate τ by finding how long it takes the current to charge to 0.632 times the final value minus the initial value.
8. *Capture* an oscilloscope screenshot showing v_{C}, v_{SENSE}, and the measurement cursors.
9. *Estimate* L_{test} using the measured τ. *Use* $R_{\text{th}} = R_{\text{sense}} + R_{\text{S}}$.
10. *Compute* the percent error of the calculated and measured capacitance.

Task 9.6.2: Light doorbell design

Deaf and hard of hearing people use light-based systems instead of audible doorbells. A light based doorbell illuminates for a period of time after the doorbell button is pushed. In this problem, we will walk through the design of an LED doorbell using a 555 timer. The timer runs in the monostable configuration, as shown in figure 9.21. Only include the 330 Ω resistor if the LED doesn't have an integrated current limiting resistor. To simplify measurements, use an on time of 5 s. It would be easy to modify the delay to something longer for a real product.

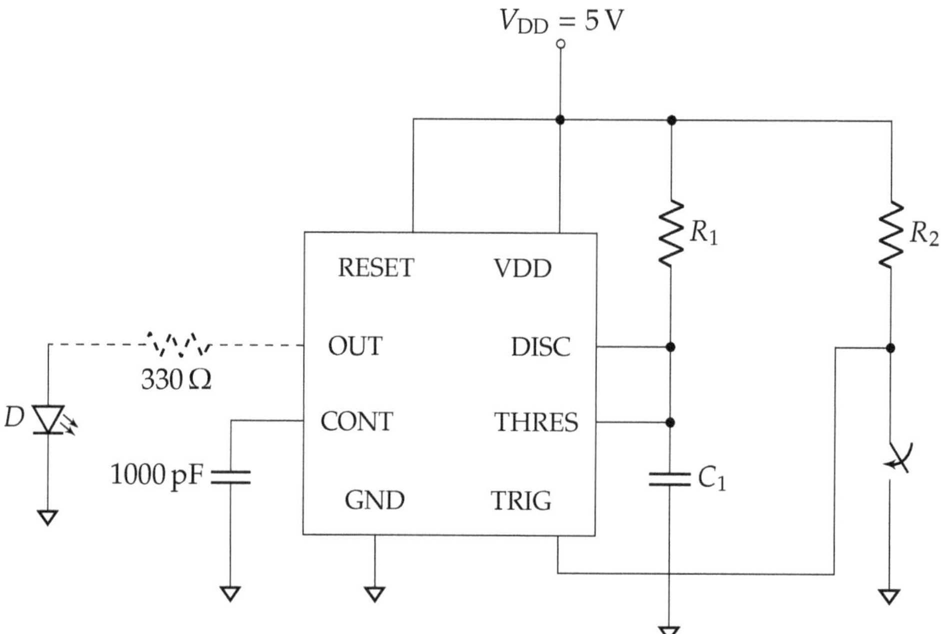

Figure 9.21: Latch circuit using a 555 timer in monostable mode.

1. *Calculate* R_1 and C_1 such that it will take 5 s to charge from 0 V to $\frac{2}{3}V_{CC}$.
2. *Construct* the circuit in figure 9.21 using an LMC555 CMOS Timer (LMC555) [3]. Add a 0.1 μF decoupling capacitor[a] between V_{DD} and ground. Use 10 kΩ for R_2.
3. *Use* single triggering mode to *capture* a printout of OUT and THRES that shows a full period. *Use* cursors to measure the width of the pulse at OUT. If you picked a large value for R_1, you may need to use x10 mode for the probe connected to THRES.
4. *Calculate* the error for the on time.

[a] Decoupling capacitors smooth the supply voltage for an IC and should be included anywhere an IC is used. The LMC555 datasheet specifically suggests 0.1 μF.

Task 9.6.3: LED strobe light design

The timer can also produce oscillating waveforms when in the astable configuration. In this problem, suppose you intend to construct a circuit that flashes an LED at a specific frequency. Only include the 330 Ω resistor if the LED doesn't have an integrated current limiting resistor.

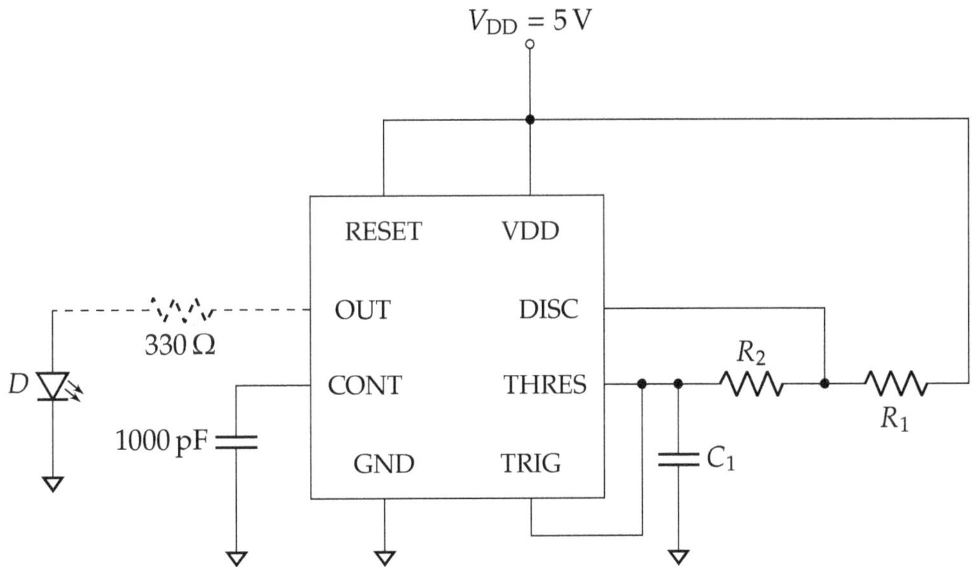

Figure 9.22: Flasher circuit using an astable mode 555 timer circuit.

1. *Construct* the circuit in figure 9.22 using the values calculated in your prelab.
2. *Capture* an oscilloscope printout showing THRES and OUT. *Include* measurements for duty cycle and frequency.
3. *Calculate* the error between the designed and measured values.
4. *Calculate* the capacitance needed to change the oscillation frequency to 10 Hz.

9.7 References

[1] Jack Ward. (2004). The 555 timer ic, [Online]. Available: `http://semiconductormuseum.com/Transistors/LectureHall/Camenzind/Camenzind_Index.htm` (visited on 02/26/2018).

[2] Paul Falstad. (2017). Circuit simulation, [Online]. Available: `https://www.electronicdevicesanddesign.com/CircuitSim/circuitjs.html` (visited on 02/27/2018).

[3] *LMC555 CMOS timer*, LMC555, SNAS558M, Texas Instruments, Jul. 2016. [Online]. Available: `http://www.ti.com/lit/ds/symlink/lmc555.pdf`.

EXPERIMENT 10

Operational Amplifier Characteristics

10.1 Application

Operational amplifiers are available in nearly unlimited varieties from countless vendors. In order to understand how to pick the correct op amp for a specific circuit, it is necessary to understand what all of the parameters on an op amp datasheet mean. A thorough understanding of op amp characteristics is invaluable to the analog circuit designer. The measurable parameters of an op amp can be used to predict how well the surrounding circuit will perform and allow for verifying that the inputs and outputs of the circuit will behave as expected.

Figure 10.1: A vaccuum tube op amp. Vaccuum tube op amps were used prior to the invention of the integrated circuit op amps we use today.

10.2 Decibels

The decibel is a unit that measures the ratio between two values on a logarithmic scale. This turns out to be useful when expressing values of extremely large dynamic range. Technically, the unit is a Bel, however, a Bel is so large that we normally speak of decibels. The definition of a *decibel* is

$$X_{\text{dB}} = 10 \log \frac{|X|}{|X_0|} \tag{10.1}$$

Note: the definition of a decibel includes absolute values on the inputs; however, more often then not those are dropped. In practice decibels are widely used with power measurements that are inherently positive. When using decibels to measure voltage and current, the AC root-mean-square (RMS) values are typically used as they are also always positive.

Where X_0 is what X is referenced to. For example, if we measured 100 mW, we could express it in decibels of milliwatts by taking

$$P_{\text{dB}} = 10 \log \frac{100\,\text{mW}}{1\,\text{mW}} = 20\,\text{dBm}$$

The unit "dBm" is used to signify that the measurement is referenced to a milliwatt.

Gains add Decibels are on a logarithmic scale and follow logarithmic arithmetic. For gain stages in series, the individual gains multiply[1] in the linear domain and *add* in the logarithmic. That is,

(1) Think two amplifiers connected in series.

$$\begin{aligned} G_{\text{dB}} &= 10 \log(G_1 G_2) \\ &= 10 \log G_1 + 10 \log G_2 \\ &= G_{1,\text{dB}} + G_{2,\text{dB}} \end{aligned}$$

where G_{dB} is the overall gain in decibels, and G_1 and G_2 are the linear power gains for each stage.

Losses subtract Similarly losses in a system *subtract* from the overall gain.

$$\begin{aligned} G_{\text{dB}} &= 10 \log\left(G_1 \frac{1}{L_2}\right) \\ &= 10 \log G_1 - 10 \log L_2 \\ &= G_{1,\text{dB}} - L_{2,\text{dB}} \end{aligned}$$

where G_{dB} is the overall gain in decibels, G_1 is the power gain of the first stage, and L_2 is the power loss[2] of the second stage.

(2) A loss, L, is the multiplicative factor of which the output is *smaller* than the input. In other words, $G = \frac{1}{L}$ represents a system *gain* of less than 1.

It is common for an engineer to say "3 dB down" or that the "gain" is −3 dB. What they mean is that the point being observed has a 3 dB attenuation, or signal reduction.

System gain It is commonplace to discuss a system's gain, but one should be careful. By convention *power* gain is used in applications where voltage or current gain may seem more relevant. That said, it is easy to translate between them, as long as the load and input resistance are equal(3) ($R_{in} = R_L$). It can be done by taking

$$G_{dB} = 10\log\frac{P_{out}}{P_{in}} = 10\log\frac{V_{out}^2/R_L}{V_{in}^2/R_{in}}$$
$$= 10\log\left(\frac{V_{out}}{V_{in}}\right)^2 = 20\log\frac{|V_{out}|}{|V_{in}|} = 20\log|A_v| \qquad (10.2)$$

(3) In a low-power signal processing system, the input resistance of each stage is typically very large and therefore negligible. For this reason, treating the input resistances of all of the stages as approximately equal is a valid assumption.

10.2.1 Gain as a function of frequency

The general study of systems with respect to frequency is a large, important topic and will be considered in detail starting with experiment 13; however, we can still appreciate the primary impact by empirically studying the effects.

For the applications we are considering, it is sufficient to say that the gain of an op amp system is low pass in nature. That is, the gain is constant for low frequencies and gradually falls off for large ones.

Measuring the system gain versus frequency is a routine task in electrical engineering. In fact, it is so important that there is an entire class of measurement tools called network analyzers that are designed to measure frequency response of systems. Crude, but effective, system gain and phase measurements can be made relatively quickly with a function generator and oscilloscope. Some oscilloscopes with built in function generators may have a rudimentary frequency response plotting tool.

A plot of the generic response for a "low-pass" system is found in figure 10.2. For reference, the gain and phase equations for a first order low-pass system are

Plots of magnitude and phase versus frequency are called "Bode" plots, named after Hendrik Wade Bode.

$$|A_v(f)| = \sqrt{\frac{A_0}{1+\left(\frac{f}{f_c}\right)^2}} \qquad \angle A_v(f) = -\arctan\frac{f}{f_c}. \qquad (10.3)$$

where A_0 is the DC gain and f_c is the so called cutoff frequency.

Notice that as the frequency increases, the gain decreases and the phase of the output relative to the input shifts. Then, after a cutoff frequency, the gain rapidly decreases by $-20\,\mathrm{dB/decade}$ and the phase continues to shift. Finally, about one additional decade(4) of frequency later, the phase settles off at a shift of -90°. The gain contains to fall off at a constant rate forever.

(4) A decade is a power of 10 times the frequency. For example, 1 kHz is 1 decade higher than 100 Hz.

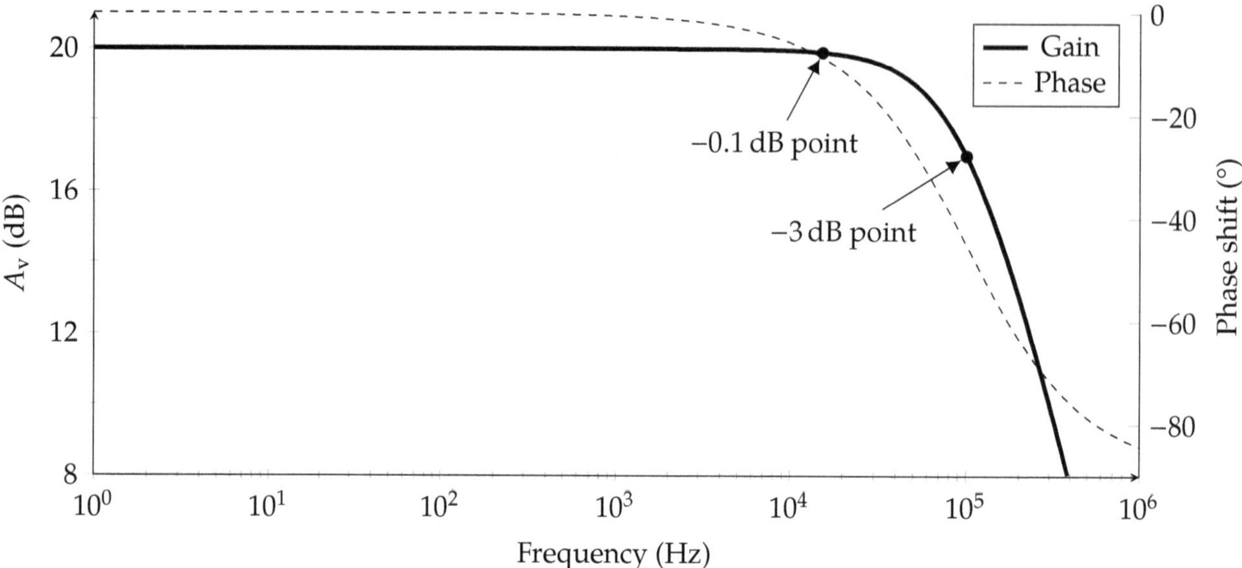

Figure 10.2: Example first order low-pass amplifier system, with −3 dB cutoff frequency, $f_{-3\,\mathrm{dB}} = 100\,\mathrm{kHz}$ and −0.1 dB cutoff frequency, $f_{-0.1\,\mathrm{dB}} = 15.262\,\mathrm{kHz}$

10.2.2 System bandwidth

From figure 10.2 it should be clear that a real system has a finite range of operation. Signals at frequencies higher than the cutoff point are significantly attenuated compared to those at lower frequencies. We say that frequency content below the cutoff point is "passed" and those above are "rejected." We refer to the maximum frequency that is passed as the system's bandwidth. There are several common definitions of bandwidth, so we will only consider the two most popular.

−3 dB bandwidth

The most commonly used definition of bandwidth is the −3 dB point. This is the frequency where the *power* gain is 3 dB lower than the passband(5). As suggested by figure 10.2, this point is not the most obvious choice to mark the end of the "flat" region. Clearly, by the −3 dB point, the system has caused significant signal attenuation and phase shifting. However, it is a convenient for several reasons:

(5) Careful– this is usually not when the power gain $G_{db} = -3\,\mathrm{dB}$! If a circuit has a passband power gain of 20 dB like in figure 10.2, then the −3 dB point is when $G_{db} = 20\,\mathrm{dB} - 3\,\mathrm{dB} = 17\,\mathrm{dB}$

1. It is approximately equal to the cutoff frequency, f_c, in equation (10.3).

2. It is the frequency at which the output signal power is approximately equal to 1/2 the input power.

To see the first, we can calculate the decibel gain of the theoretical response (equation (10.3)) and evaluate it such that $f = f_c$:

$$\begin{aligned} G_{-3\,\text{dB}} &= 20 \log \left. \sqrt{\frac{1}{1 + \left(\frac{f}{f_c}\right)^2}} \right|_{f=f_c} \\ &= 20 \log \sqrt{\frac{1}{2}} = -3.0103\,\text{dB} \approx -3\,\text{dB} \end{aligned}$$

Notice that this point is not perfectly −3 dB; however, it is so close that in practice we just use −3 dB.

To prove the second property of the −3 dB point, we notice that

$$20 \log \sqrt{\frac{1}{2}} = 10 \log \frac{1}{2}\ \text{dB}$$

and recall that the definition of system gain in decibels is the ratio of output power to input power. In this case, the ratio is 1/2 as predicted.

Working with voltage gains

Power gain is not the most useful tool in a voltage processing system,[6] even if it is the status quo. Thankfully, we can convert the −3 dB gain rule to a voltage gain by calculating when the output power is half of the passband gain. We once again need to assume that $R_{\text{in}} = R_{\text{L}}$.

[6] For example, an op amp circuit.

$$\begin{aligned} \left.\frac{P_{\text{out}}}{P_{\text{in}}}\right|_{-3\,\text{dB}} &= \frac{1}{2} \left.\frac{P_{\text{out}}}{P_{\text{in}}}\right|_{\text{passband}} \\ \left.\frac{V_{\text{out}}^2/R_{\text{L}}}{V_{\text{in}}^2/R_{\text{in}}}\right|_{-3\,\text{dB}} &= \frac{1}{2} \left.\frac{V_{\text{out}}^2/R_{\text{L}}}{V_{\text{in}}^2/R_{\text{in}}}\right|_{\text{passband}} \\ \left.A_{\text{v}}^2\right|_{-3\,\text{dB}} &= \frac{1}{2} \left.A_{\text{v}}^2\right|_{\text{passband}} \\ A_{\text{v,-3dB}} &= \frac{1}{\sqrt{2}} A_{\text{v,passband}} \approx 0.707 \cdot A_{\text{v,passband}} \end{aligned}$$

When the voltage gain reduces to approximate 0.707 times the ideal passband gain of the system, it has reached the −3 dB point. Once again, it is important to note that the −3 dB point is *not* when the gain is equal to 0.707.

−0.1 dB bandwidth

If a design requires very little loss of phase difference between input and output in the passband, then setting the system bandwidth by the −0.1 dB point may be a reasonable alternative. This bandwidth allows for only a very small change in the gain magnitude and moderate phase shifts.

As with the −3 dB bandwidth, we can find the voltage gain that is equivalent to the −0.1 dB point by taking

$$20\log|A_{\text{passband}}| - 0.1\,\text{dB} = 20\log|A_{-0.1\,\text{dB}}|$$
$$|A_{-0.1\,\text{dB}}| = 10^{\frac{-0.1}{20}}\left|A_{\text{passband}}\right| \approx 0.9886\left|A_{\text{passband}}\right|$$

10.3 The real op amp

In experiment 8: "Operational Amplifiers: Basic Circuits" we studied the op amp in an ideal sense. That is, we used the ideal virtual short circuit analysis that effectively assumes all op amps have

1. infinite input resistance,
2. zero input current,
3. infinite gain, and
4. zero output resistance.

However, if these assumptions were truly the case, then there would be no need for multiple op amp designs in the marketplace. Recall that in our short coverage of op amps, we have already used several models, which is evidence (even if only anecdotal) that there must be some non-ideal effects.

Thankfully, amongst the many thousands of op amps on the market, one can usually find some that realize the relevant properties so well that the device can be considered ideal for the problem at hand. The trick is selecting the correct op amp for the job. To do that, one must study how the non-ideal characteristics manifest, the relevant datasheet metrics, and finally the impacts on a given circuit topology.

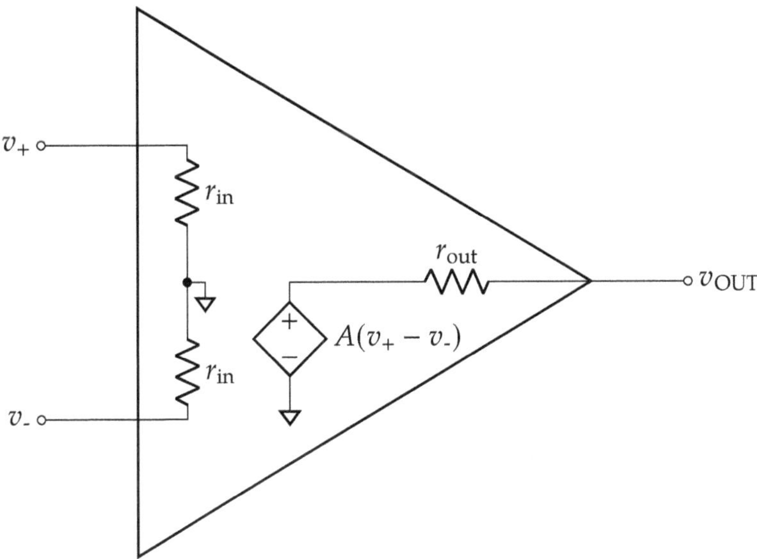

Figure 10.3: A model that better reflects real op amps, where r_{in} is the input resistance, A is the op amp gain, and r_{out} is the output resistance.

Figure 10.3 is the op amp circuit model from experiment 8 updated to reflect the real nature of both inputs and the output. We previously ignored the input resistance, output resistance, and finite gain of a real op amp. In most applications, the ideal model is adequate for the design of an op amp based system, but the limitations of an op amp become apparent in low power, low voltage, high frequency, or high gain circuits. We will work through most of the parametrics present in an op amp's datasheet in order to aid with op amp selection in your future circuit designs.

Finite gain and bandwidth

The gain of an op amp, known as the *open loop gain*, is very large. Ideally, it would be infinite; however, in practice, it is usually $\gg 200,000$. As long as the open loop gain stays much larger than the designed gain of the circuit, then the op amp will function as expected. Recall that a real op amp acts like a low-pass filter which reduces its gain over frequency. Therefore, you should expect an op amp to stop performing as well at high frequency. The gain-bandwidth product is helpful in understanding what frequency is too high.

Gain-bandwidth product (*GBW*)

The gain-bandwidth product (*GBW*) is as simple as the name implies: $GBW = \text{gain} \times \text{bandwidth}$. The surprising thing is that the product is approximately a constant! In other words, if you know the *GBW* value from an op amp's datasheet[(7)], then one can compute the maximum bandwidth in which the op amp will function for a certain gain and vice versa. Most manufactures cite the product using the 3 dB bandwidth. Gain-bandwidth is often one of the first considerations when picking an op amp.

[(7)] Some datasheets call this the "unity gain bandwidth," which is the frequency at which the op amp's gain is equal to 1. Of course, this is the same metric as *GBW*.

Finite input resistance

Ideally an op amp will have an infinite input resistance but practically there is a limit. The good thing is that modern FET-based op amps can realize such large resistances ($> 10^{10}\ \Omega$) that they are effectively infinite. Even BJT-based designs should have input resistances in the mega ohm range.

Input bias current

The input bias current is ideally zero. For a real op amp it cannot be perfectly zero because the internal amplifier must draw some amount of current to turn on. The bias current is approximately DC and, while extremely small in modern op amps (nano to pico amps), can not always be treated as zero.

For example, if the input of a non-inverting op amp circuit is capacitively coupled as illustrated in figure 10.4, then the input bias current from the non-inverting terminal will slowly charge the coupling capacitor until it charges to nearly the supply. In turn, this causes the op amp output to saturate. In order to solve this problem, a resistor must be connected between the capacitor and the positive input; however, the added resistor is in parallel with the massively large input resistance of the op amp, so the overall input resistance of the circuit as a whole decreases.

Figure 10.4: Capacitively coupled op amp input – don't do this!

Input offset voltage

The input offset voltage is the voltage that must be externally applied across the device's inputs[(8)] to give a zero volt output. By design the offset voltage should be zero; however, in practice the input stage of an op amp is slightly unbalanced. This small imbalance appears as parasitic input to the op amp in addition to the feedback voltage. An example of input offset is shown in figure 10.5. The output of the circuit should be much greater than 0, but the input offset cancels out the 1 mV

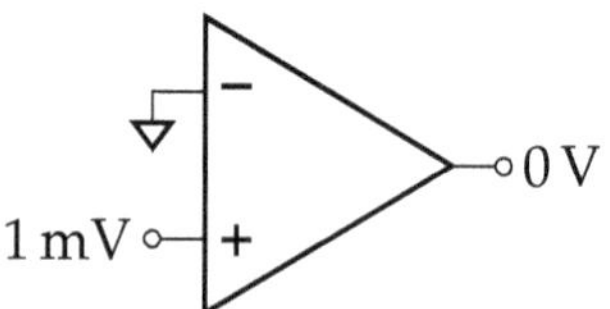

Figure 10.5: Behavior of input offset voltage

[(8)] That is, the voltage at each input is slightly different.

from the positive input. In effect, the dependent source in figure 10.3 actually follows the equation:

$$A(v_+ - v_- \pm v_{\text{off}})$$

Where v_{off} is the input offset voltage. The datasheet for an op amp will usually report the worst case input offset voltage *magnitude*, so the sign can be positive or negative.

Input offset current

The input offset current is the amount of mismatch between the input bias current at each input terminal. That is, $i_+ - i_-$. This can cause a mismatch in the voltage at the input terminals via the surrounding resistor network and result in a DC error on the output. In modern op amps this mismatch is typically so small that it can normally be ignored unless the feedback has a particularly high resistance.

Slew rate

Slew rate is defined as the change of electrical quantity per unit of time. In the context of op amps, it is defined as the maximum rate of change of voltage at the output. The rate is usually defined as volts per microsecond.

If the output is a time varying signal, then one can take the derivative with respect to time to determine the slew rate of the signal. Usually, we are interested in keeping the maximum slew rate appreciably below the op amp's stated specification.

Input common-mode voltage

As we explored in experiment 8, one must ensure that the common mode voltage, $v_{\text{CM}} = \frac{v_+ + v_-}{2}$, stays within the op amps required range at all times. This is particularly an issue for non-inverting amplifiers and single sided designs. The input common mode voltage is an incredibly important aspect of picking an op amp.

Note: It is important to ensure that the input voltages remain within the absolute maximum voltage specifications of an op amp and satisfy the input common-mode voltage requirements.

Common-mode gain (common-mode rejection ratio)

The common-mode gain of an op amp is very small. That is, it effectively eliminates any voltage that is common to both the inverting and non-inverting terminals. In other words, it is a good differential amplifier. For example, the op amp in figure 10.6 should produce an output of $v_{\text{OUT}} = 0\,\text{V}$ no matter what v_{IN} is, but a non-ideal op amp will pass some of the common-mode input to the output.

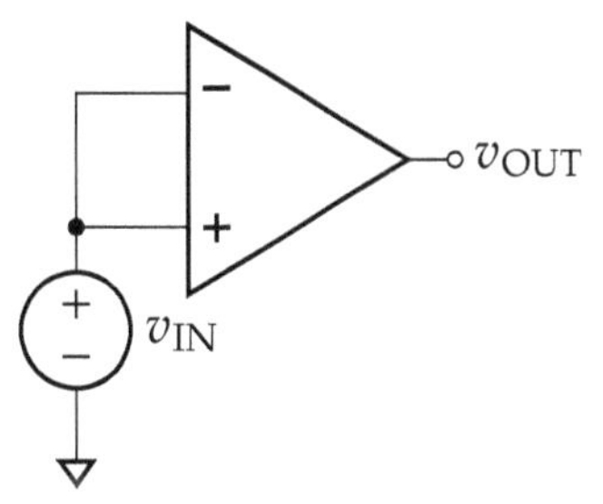

Figure 10.6: Common mode gain circuit

To measure how well it suppresses the common-mode voltage, datasheets typically cite a common-mode rejection ratio. The rejection ratio is the ratio between the common-mode gain and the open loop gain.

Power supply rejection

The power supply rejection ratio (PSRR) is the ratio between the change in the supply voltage of the op-amp to the change in output voltage.

$$PSRR = \frac{\text{change in } V_{CC}}{\text{change in } v_{OUT}}$$

10.4 Computing the input resistance of op-amp circuits

While there are many non idealities to consider in an op amp, the surrounding circuitry still dominates the behavior of an op amp circuit in most cases, especially with regards to input and output resistance. While the inputs to an op amp often behave like ideal open circuits, the input resistance of an op amp circuit is usually set by the feedback resistors.

Example 10.4.1: Finding the input resistance of an inverting amplifier

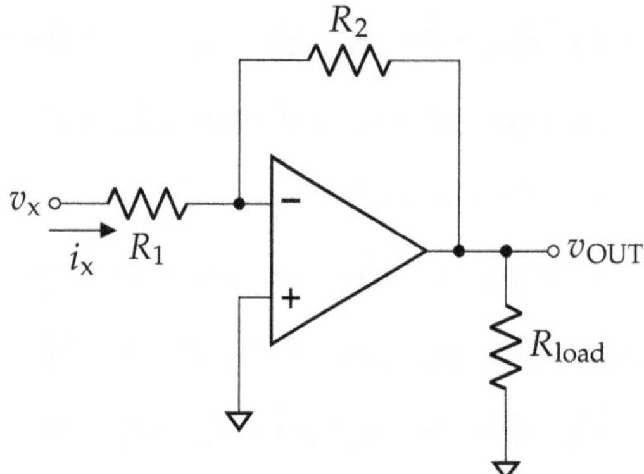

Figure 10.7: An inverting amplifier circuit.

To measure the input resistance, we apply a theoretical test source, v_X, at the input and ask what current, i_X, flows from it. Once we know the current in terms of v_X, we can use Ohm's law to compute the effective input resistance.

Using the virtual circuit analysis, assuming the op amp is functioning close to ideal, we say that $v_- = v_+ = 0\,\text{V}$. Therefore, the voltage drop across R_1 is v_X and we can compute

i_x.

$$i_x = \frac{v_x - v_-}{R_1} = \frac{v_x - 0}{R_1} = \frac{v_x}{R_1}$$

Then, using Ohm's law, the input resistance is found to be:

$$\begin{aligned} R_{in} &= \frac{v_x}{i_i} \\ &= \frac{v_x}{\frac{v_x}{R_1}} = R_1 \end{aligned}$$

In other words, given that the op amp is functioning correctly, the input resistance of a non-inverting amplifier is equal to the input resistor, *not the incredibly large input resistance of the op amp itself.*

Example 10.4.2: Finding the input resistance of a non-inverting amplifier

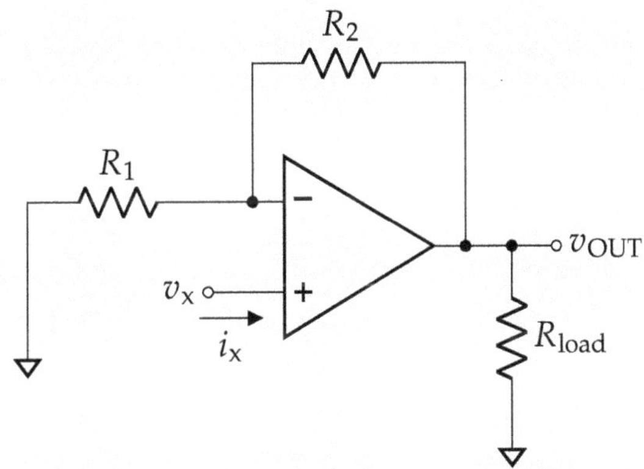

Figure 10.8: A non-inverting amplifier circuit.

Looking back to the real model of an op amp in figure 10.3, notice that the plus input terminal looks directly into the op amps differential resistance. Therefore, the test source looks into this resistance in series with whatever is present at the minus terminal, even accounting for feedback action. Regardless of the path to ground available at the minus terminal, the input resistance is at least equal to r_{in}.

For most op amps the differential resistance is at least megaohms, and for JFET or FET based inputs, it can be as high or higher than 10^{12} or 10^{13} Ohms. In other words, it is so large that it is difficult to measure. It is hard to imagine scenarios in which this input resistance is not sufficient to avoid loading.

10.5 Prelab

Task 10.5.1: Prelab questions

Please complete the following tasks

1. *Show* that the voltage gain in decibels of figure 10.9 is

$$G_{\text{db}} = 20\log\frac{R_2}{R_1} + 20\log\frac{R_4}{R_3}$$

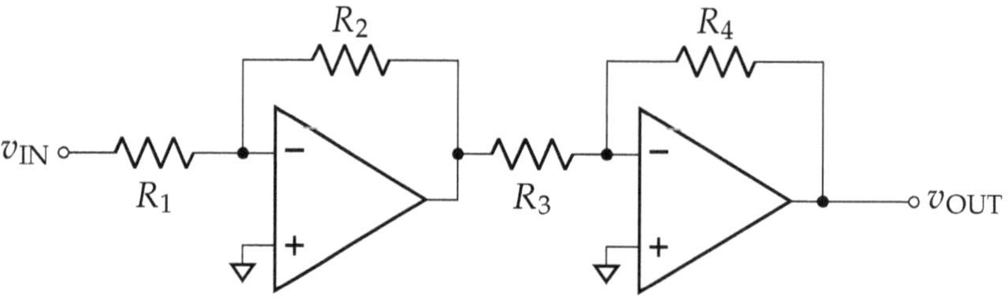

Figure 10.9: Cascaded op amp circuit

2. *Derive* the input and output relationship for the differential amplifier in task 10.6.3 figure 10.14.[a]
   a) Why is this amplifier called a *differential amplifier*?
   b) What occurs when $R_1 = R_2 = R_3 = R_4$?

3. *Look up* the following op amp characteristics in the UA741, LF356N, and LM324N datasheets. Some of these parameters must be estimated from plots or may not be reported in some datasheets. Use room temperature and a supply voltage of ±15 V when multiple specifications are reported.
   a) Gain bandwidth product or unity gain bandwidth
   b) Input resistance
   c) Input bias current
   d) Input offset voltage
   e) Common mode input voltage range
   f) Output voltage swing with $R_L = 2\,\text{k}\Omega$
   g) Output current (sourcing and sinking)

4. *Lookup* each part's quantity 1000 price. How do the characteristics follow with the components price?

[a] Hint: Use superposition to find the output due to each input and then combine.

Task 10.5.2: Prelab task: Slew rate

You can use the Analog Discovery 2 to complete this task.

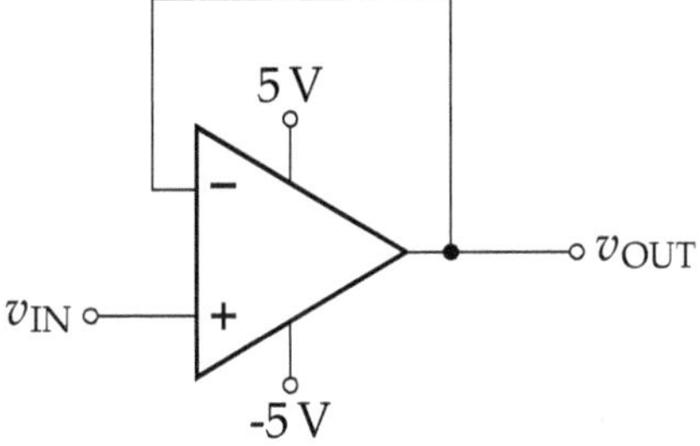

Figure 10.10: Slew rate test circuit.

Use the LM324N to complete the following tasks. **You must install decoupling capacitors.**

1. *Apply* an input square wave at 10 kHz such that the output is 6 $V_{p\text{-}p}$ with no offset.
2. Does the output look correct for a buffer circuit?
3. *Zoom* into the transition on the output. This signal is slew rate limited. *Measure* the slew rate.
4. *Compute* the theoretical slew rate for $1.5 \sin(2\pi f t)$ in terms of f.
5. *Calculate* the frequency, f_{slew}, in which the signal's maximum slew rate is equal to the measured op amp slew rate.
6. *Apply* the signal and slightly adjust the frequency to 0.8, 1.0, and 1.2 times f_{slew}. *Capture* a screenshot for each case. Is the output distorted for any of the cases? Are you within the gain bandwidth product limit?

10.6 Tasks

Task 10.6.1: Op-amp gain bandwidth measurement

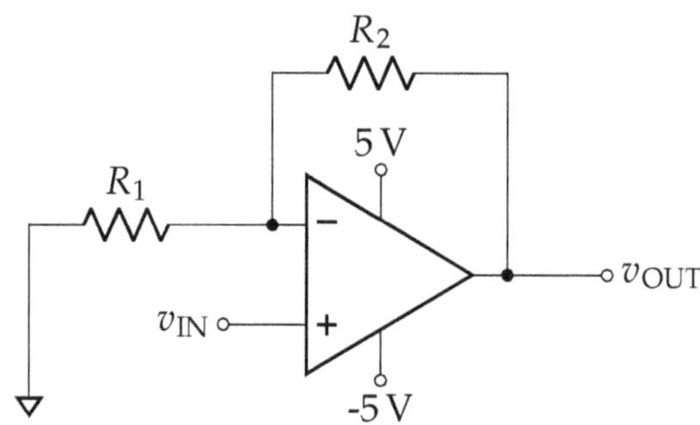

Figure 10.11: A non-inverting amplifier circuit.

Use the LM324N to complete the following tasks. **You must install decoupling capacitors.**

1. For $R_1 = 10\,\text{k}\Omega$ and $R_2 = 22\,\text{k}\Omega$, complete:
   a) *Determine* the expected gain of the amplifier.
   b) *Use* a 200 $\text{mV}_{\text{p-p}}$ sinusoidal input.
   c) *Step* the frequency of the input from 1 kHz to 1 MHz by powers of ten. *Record* both the input and output **AC RMS** value and phase.[a]
   d) *Plot* gain and phase versus frequency.
   e) *Estimate* the −0.1 dB and −3 dB points.
   f) *Find* them precisely by adjusting the function generators frequency and making scope measurements.
   g) *Compute* the gain bandwidth product using the −3 dB bandwidth.
2. *Repeat* step 1 with $R_2 = 33\,\text{k}\Omega$.
3. What is the relationship of the gain bandwidth product between the two gains?
4. How do the measured gain bandwidth products compare to the datasheet claims?
5. How can you use the gain bandwidth product to help select an appropriate op amp in future designs?

[a] Do not assume that the input voltage is equal to what your function generator says, measure it!

Task 10.6.2: Input bias current

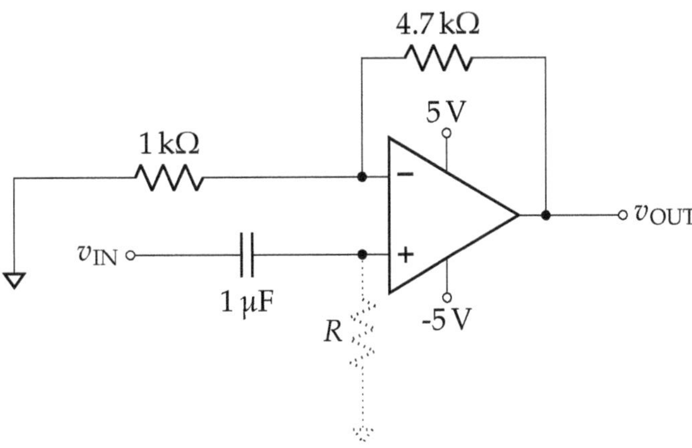

Figure 10.12: An amplifier circuit for investigating the effect of input bias current.

1. *Construct* the circuit in figure 10.12 with $R = 1\,\mathrm{M\Omega}$ and a LM324N op-amp [1]. **Use decoupling capacitors.**
2. *Apply* $v_{IN} = 0.2\sin(2\pi 10^3 t) + 1$ and *measure* v_{OUT}.
3. Does the output match your expectations?
4. *Remove* the dashed resistor and observe the output. *Describe* what happens after the resistor is removed.
5. If the R is not included, and C is very large, what would the circuit ultimately do? Would this be easy to miss during normal debugging?

Task 10.6.3: Signal restoration using op-amp circuits

Use the LM324N to complete the following tasks. **You must install decoupling capacitors.**

The Controller Area Network (CAN) bus is a prevalent network standard that allows microcontrollers and other devices to communicate with each other in electrically noisy environments. It is widely used in road vehicles, agricultural machinery, and watercrafts[a].

The CAN bus consists of two wires, CAN high and CAN low. To send a "0" bit, the CAN high wire is driven to 3.5 V and the CAN low wire to 1.5 V. To send a "1" bit, both CAN high and low are driven to 2.5 V. An example noisless sequence of CAN high and CAN low waveforms for the sequence 1010111101 is shown in figure 10.13. In this case, each bit is transmitted for 0.1 seconds.

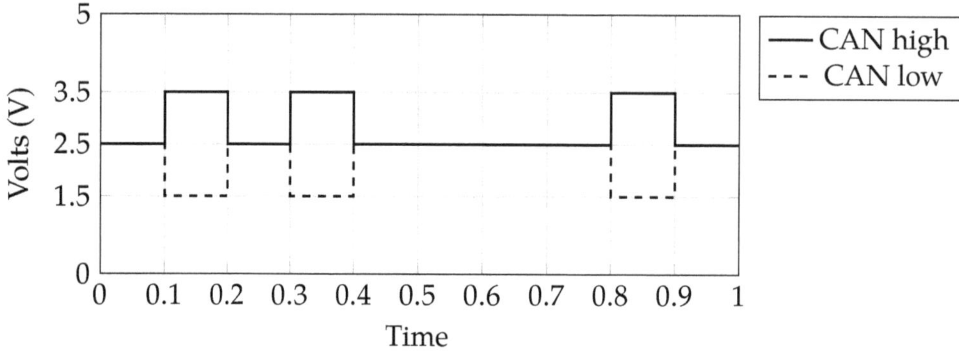

Figure 10.13: A sequence of CAN signals. For demonstration purposes, if we define time interval of 0.1 seconds per bit, then we can decode the waveform to be 1010111101 in binary.

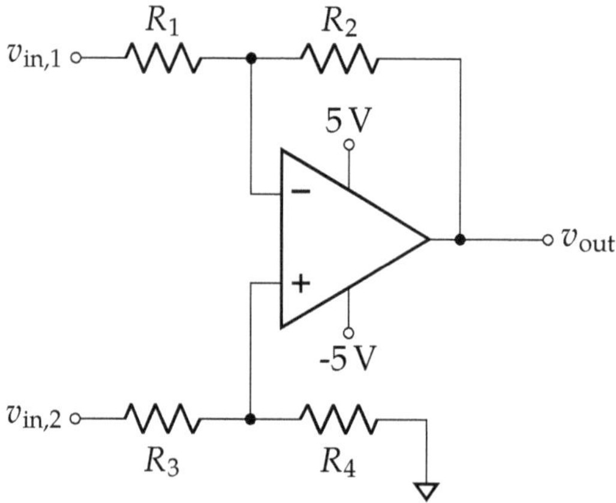

Figure 10.14: A differential amplifier circuit.

In this task, a series of CAN signals are simulated with the function generator. We assume that during the signals transmission, it is corrupted by random electromagnetic noise. However, because both the CAN high and low wires were routed through the vehicle next to each other the whole way, the noise picked up by both are approximately equal.

Your task is to remove the noise and retrieve the level shifted[b] bits with voltage level 0 V and 5 V coresponding to logic level 0 and 1, respectively.

1. *Run* the provided software to generate the CAN signal. Function generator output 1 is CAN high and output 2 is CAN low.
2. *Capture* both CAN high and low on the scope. Can you visually decode the bits with the noise?
3. *Design* a circuit that removes the noise and level shifts the waveform to 0 V and 5 V bit pulses. The differential amplifier circuit in figure 10.14 may be particularly useful. You may use more than one op amp if needed. You may want to use a comparator to invert the output and get a clean 0 V and 5 V.
4. *Capture* the result and visually decode the first 8 bits given that each bit is 1.25 ms long. *Use* the function generator's sync output to determine the start of the bit sequence.

[a] Note: both agricultural machinery and watercrafts applications use higher level protocols that are based on CAN protocol; the former uses SAE J1939 and the latter uses NMEA 2000.

[b] A level shifter is a circuit that translates signals from one logic level to another, allowing compatiblity between different circuit components.

10.7 References

[1] *LMx24-N, LM2902-N low-power, quad-operational amplifiers*, LM324N, SNOSC16D, Texas Instruments Inc., Jan. 2015. [Online]. Available: `http://www.ti.com/lit/ds/symlink/lm324-n.pdf`.

EXPERIMENT 11

RLC Step Response

11.1 Application

Parallel and series RLC circuits finds their use in many filter and oscillator designs. The resonant tank formed by an inductor and capacitor is also used widely in analog radios. A tunable capacitor in tandem with an inductor can be used to adjust the resonant frequency of a circuit to the desired radio station. RLC circuits are also used to model many non-idealities of wires and circuit board traces, so analysis of RLC circuit behavior can enable predicting non-ideal behavior in high frequency digital and analog systems.

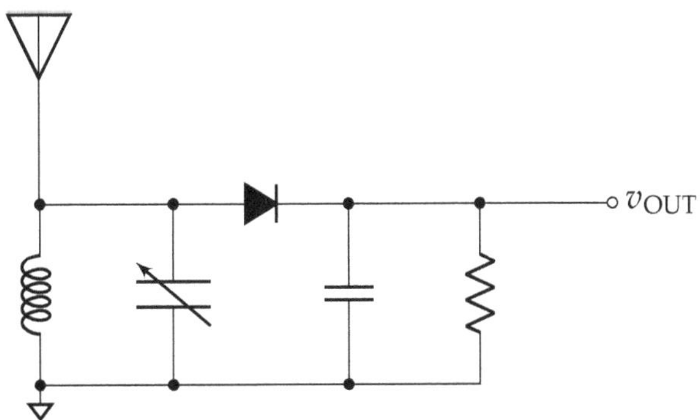

Figure 11.1: A simple AM radio detector using a resonant LC circuit

11.2 Second order systems

Resistor-Inductor-Capacitor (RLC) circuits are linear second order systems because they are fully described by linear second order differential equations. To successfully study these type of circuits, one must find the solution to the generalized second order system equation shown in equation (11.1). Note that $f(t)$ is the forcing function and is the input to the system. $x(t)$ is the output and is typically a current or voltage signal.

$$\frac{\mathrm{d}^2 x(t)}{\mathrm{d}t^2} + b\frac{\mathrm{d}x(t)}{\mathrm{d}t} + cx(t) = f(t) \tag{11.1}$$

11.2.1 Unforced solution (homogeneous)

We begin our analysis with the homogeneous case, $f(t) = 0$. In this case, the system is un-forced and becomes

$$\frac{\mathrm{d}^2 x_0(t)}{\mathrm{d}t^2} + b\frac{\mathrm{d}x_0(t)}{\mathrm{d}t} + cx_0(t) = 0. \tag{11.2}$$

The solution, $x_0(t)$, takes the form Ke^{st} where K is a constant. It may not be immediately obvious that the solution is an exponential. Convince yourself that to satisfy both sides of equation (11.2), the solution must have a derivative equal to a constant multiple of itself.[(1)]

[(1)] If these concepts are unclear to you or if you prefer a more rigorous proof, we suggest that you review homogeneous solutions in a differential equations textbook before you continue.

To find the specific solution, plug the general solution into equation (11.2) and solve.

$$\begin{aligned} \frac{\mathrm{d}^2}{\mathrm{d}t^2}\left(Ke^{st}\right) + b\frac{\mathrm{d}}{\mathrm{d}t}\left(Ke^{st}\right) + ce^{st} &= 0 \\ s^2Ke^{st} + bsKe^{st} + cKe^{st} &= 0 \\ Ke^{st}\left(s^2 + bs + c\right) &= 0 \end{aligned}$$

Ignoring the trivial solution ($K = 0$), we can deduce that

$$s^2 + bs + c = 0 \tag{11.3}$$

must be true. Equation (11.3) is known as the characteristic equation and has two solutions described by the quadratic formula:

$$s = \frac{-b \pm \sqrt{b^2 - 4c}}{2}$$

Therefore, a system described by equation (11.2) has two solutions

$$x_{0,1}(t) = K_1e^{s_1t} \qquad \text{and} \qquad x_{0,2}(t) = K_2e^{s_2t}$$

where $x_{0,i}(t)$ is the solution corresponding to the i-th root

$$s_1 = \frac{-b + \sqrt{b^2 - 4c}}{2} \quad \text{and} \quad s_2 = \frac{-b - \sqrt{b^2 - 4c}}{2}$$

This differential equation is linear, so the overall solution is the sum of the two individual ones. More generally, the overall answer space is spanned by the two.

In summary, the system's homogeneous solution, $x_0(t)$, is given by

$$x_0(t) = x_{0,1}(t) + x_{0,2}(t) = K_1 e^{s_1 t} + K_2 e^{s_2 t}$$

where K_1 and K_2 are determined by the system's initial conditions.

11.2.2 Forced solution (nonhomogeneous)

For the purposes of this experiment, we limit our attention to nonhomogeneous case with *constant* force, $f(t) = F$. In this case, the system becomes

$$\frac{\mathrm{d}^2 x_0(t)}{\mathrm{d}t^2} + b\frac{\mathrm{d}x_0(t)}{\mathrm{d}t} + c x_0(t) = F. \tag{11.4}$$

The homogeneous solution, $x_0(t)$, no longer holds because the right hand side of the system has changed.(2) We must try a new form. Since the system is forced with a constant, let us assume that the solution is the same, but with a constant offset, G.

(2) This can be verified by plugging $x_0(t)$ into equation (11.4) and identifying the disagreement.

$$x_F(t) = x_0(t) + G = K_1 e^{s_1 t} + K_2 e^{s_2 t} + G \tag{11.5}$$

Substituting equation (11.5) into equation (11.4), we find that

$$\frac{\mathrm{d}^2}{\mathrm{d}t^2}(x_0(t) + G) + b\frac{\mathrm{d}}{\mathrm{d}t}(x_0(t) + G) + c(x_0(t) + G) = F$$

$$\frac{\mathrm{d}^2 x_0(t)}{\mathrm{d}t^2} + \cancelto{0}{\frac{\mathrm{d}^2 G}{\mathrm{d}t^2}} + b\frac{\mathrm{d}x_0(t)}{\mathrm{d}t} + b\cancelto{0}{\frac{\mathrm{d}G}{\mathrm{d}t}} + c x_0(t) + cG = F$$

$$\cancelto{0}{\frac{\mathrm{d}^2 x_0(t)}{\mathrm{d}t^2} + b\frac{\mathrm{d}x_0(t)}{\mathrm{d}t} + c x_0(t)} + cG = F$$

$$cG = F$$

$$G = \frac{F}{c}$$

Therefore, the nonhomogeneous constant force solution, sometimes called the step response, is given by equation (11.6)

$$x_F(t) = K_1 e^{s_1 t} + K_2 e^{s_2 t} + \frac{F}{c} \tag{11.6}$$

where K_1 and K_2 are determined by the system's initial conditions and s_1 and s_2 are the same as the homogeneous case.

Constant forcing generalizes the homogeneous solution

For this experiment the constant forcing response of equation (11.6) is sufficient because we only consider *constant* input systems.(3) The homogeneous solution is just a special case of the nonhomogeneous one, where $F = 0$. That said, if a system were to have a forcing function that is not constant, then the solution changes.

(3) Even if the constant input is sometimes just zero.

11.2.3 Response types

The homogeneous portion of the response is defined by two roots, s_1 and s_2. Depending on the differential equation's coefficients, they can be either real or complex. This leads to three classes of solutions: overdamped (real and distinct roots), critically damped (real and repeated roots), and underdamped (complex roots).

Overdamped (real and distinct roots)

The overdamped case occurs when both of the roots of the homogeneous solution are real and distinct from each other. That is,

$$b^2 - 4c > 0.$$

The solution, $x_F(t)$, is identical to equation (11.6)

$$x_F(t) = K_1 e^{s_1 t} + K_2 e^{s_2 t} + \frac{F}{c} \tag{11.7}$$

where $s_1 = \frac{-b+\sqrt{b^2-4c}}{2}$ and $s_1 = \frac{-b-\sqrt{b^2-4c}}{2}$. K_1 and K_2 are found by equating the initial conditions $x(0)$ and $x'(0)$ with $x_F(t)$ and its derivative at $t = 0$, respectively. The resulting relationships are

$$\begin{aligned} x(0) &= K_1 + K_2 + \frac{F}{c} \\ x'(0) &= s_1 K_1 + s_2 K_2 \end{aligned}$$

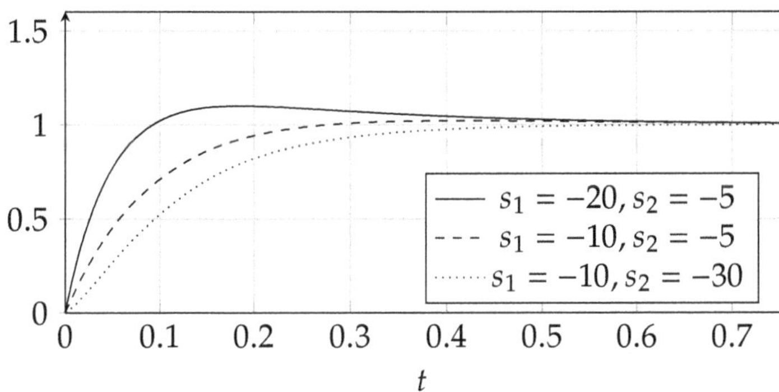

Figure 11.2: Normalized overdamped responses for constant input and initial conditions of $x(0) = 0$. In other words, the RLC overdamped step response.

Example overdamped responses for constant inputs can be found in figure 11.2. Notice how the responses smoothly rise to the final value with no oscillations with a speed based on the exponent. Interestingly, even though the system damps the response, it is still possible to overshoot the final value before settling down.

Critically damped (repeated real roots)

The critically damped case occurs when both roots are real and equal to each other. That is,

$$b^2 - 4c = 0.$$

The coincident roots causes a small hiccup in applying equation (11.6) directly. Repeated roots actually have a second solution of the form form Kte^{st}. Therefore, the critically damped response is

$$x_F(t) = (K_1 + tK_2)e^{st} + \frac{F}{c}, \tag{11.8}$$

where $s = -\frac{b}{2}$. K_1 and K_2 are found by equating the initial conditions $x(0)$ and $x'(0)$ with $x_F(t)$ and its derivative at $t = 0$, respectively. This resulting relationships are

$$x(0) = K_1 + \frac{F}{c}$$
$$x'(0) = sK_1 + K_2$$

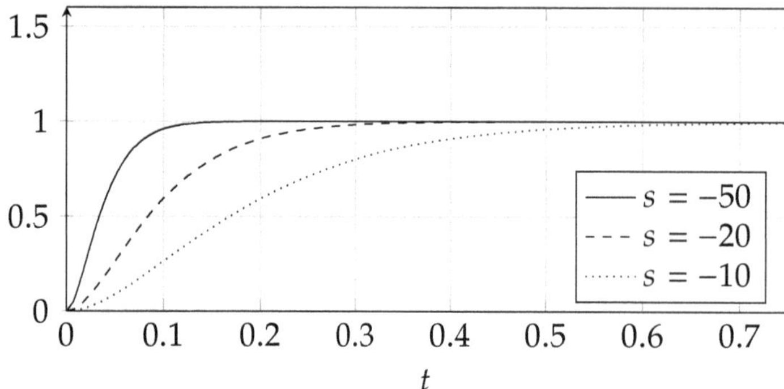

Figure 11.3: Normalized critically damped responses for constant input and initial conditions $x(0) = 0$ and $x'(0) = 0$. In other words, the RLC critically damped step response.

Examples of critically damped responses given constant input can be found in figure 11.3. The critically damped point is directly between the over and underdamped cases. At this point, the response is as fast as possible without oscillation. Practically, a system can not be built to consistently operate in the critically damped region. Component tolerances and random noise will push the response into the over or underdamped regions.

Underdamped (complex roots)

The underdamped case occurs when the roots are complex(4) conjugate roots. That is,

(4) To avoid confusion with current, we use $j = \sqrt{-1}$ rather than i.

$$b^2 - 4c < 0.$$

In this case the roots are

$$s_1 = \frac{-b + j\sqrt{4c - b^2}}{2} \qquad s_2 = \frac{-b - j\sqrt{4c - b^2}}{2}$$

For future convenience, we separate the roots into their real and imaginary parts such that

$$s = -\frac{b}{2} \pm j\frac{\sqrt{4c - b^2}}{2} = -\sigma \pm j\omega_d$$

Where σ and ω_d can be defined:

$$\sigma = \frac{b}{2} \qquad \omega_d = \frac{\sqrt{4c - b^2}}{2}$$

The solution, $x_F(t)$, is found by plugging the roots into equation (11.6)

$$\begin{aligned} x_0(t) &= K_1 e^{(-\sigma+j\omega_d)t} + K_2 e^{(-\sigma-j\omega_d)t} + \frac{F}{c} \\ &= e^{-\sigma t}\left(K_1 e^{j\omega_d t} + K_2 e^{-j\omega_d t}\right) + \frac{F}{c} \\ &= e^{-\sigma t}(A\cos\omega_d t + B\sin\omega_d t) + \frac{F}{c} \end{aligned}$$

where the final equality is by Euler's formula.[5] The coefficients are $A = K_1 + K_2$ and $B = j(K_1 - K_2)$.

[5] $e^{j\theta} = \cos(\theta) + j\sin(\theta)$

In our study of second order systems, we are modeling the response of voltages and currents, so it may surprise you that B involves an imaginary number. All is well, it just means is that realizable RLC circuits have a K_1 and K_2 such that A and B end up real. Following this line of thinking, we no longer attempt to find K_1 and K_2, but rather focus on the real valued A and B. With that, the solution has two convenient forms:

$$\begin{aligned} x_F(t) &= e^{-\sigma t}(A\cos\omega_d t + B\sin\omega_d t) + \frac{F}{c} \\ &= Re^{-\sigma t}\cos(\omega_d t + \phi) + \frac{F}{c} \end{aligned} \tag{11.9}$$

where $R = \sqrt{A^2 + B^2}$ and $\phi = \arctan(B/A)$. A and B are found by equating the initial conditions $x(0)$ and $x'(0)$ with $x_F(t)$ and its derivative at $t = 0$, respectively. The resulting relationships are

$$\begin{aligned} x(0) &= A + \frac{F}{c} \\ x'(0) &= B\omega_d - A\sigma \end{aligned}$$

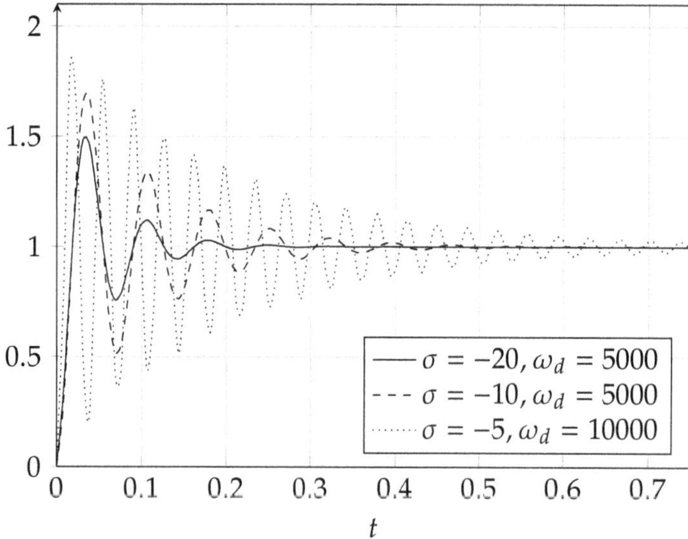

Figure 11.4: Normalized underdamped responses for constant input and initial conditions $x(0) = 0$ and $x'(0) = 0$. In other words, the RLC underdamped step response.

We can see from the simplified version of equation (11.9) that we expect the response to be sinusoidal with decaying amplitude. Figure 11.4 presents some example constant input and underdamped responses. The underdamped response oscillates around the final value and can significantly overshoot the final value. However, it has the fastest rise time of all three cases.

11.3 Second order systems for circuits

The system equations for an RLC circuit can be found using Kirchhoff's law and the fundamental component equations. The system responses are found by determining which case it is and applying equations (11.7) to (11.9) as needed.

Examples 11.3.1 and 11.3.2 will demonstrate computing a second order circuit's system equation as well as predicting the responses.

Example 11.3.1: Solving a parallel RLC circuit in a underdamped response

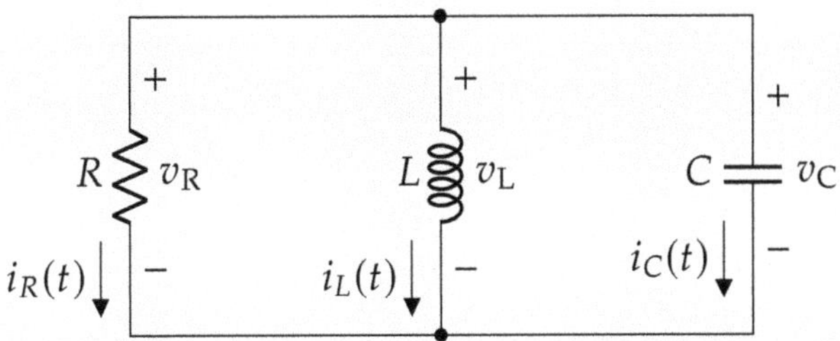

Figure 11.5: A parallel RLC circuit.

Determine the voltage, $v_c(t)$. Let $L = 10\,\mu\text{H}$, $C = 40\,\mu\text{F}$, and $R = 20\,\Omega$. The initial conditions are $v_c(0) = 10\,\text{V}$ and $i_C(0) = 0\,\text{A}$.

Find the system equation We begin by finding the system equation. Notice that all of the components in figure 11.5 are in parallel, so

$$v(t) = v_R(t) = v_L(t) = v_C(t)$$

Using Kirchhoff's current law, we have

$$\begin{aligned} i_R(t) + i_L(t) + i_C(t) &= 0 \\ \frac{v(t)}{R} + \frac{1}{L}\int v(t)\,dt + C\frac{dv(t)}{dt} &= 0 \end{aligned}$$

Now we take the derivative of both sides to eliminate the integral.

$$C\frac{d^2v(t)}{dt^2} + \frac{1}{R}\frac{dv(t)}{dt} + \frac{1}{L}v(t) = 0$$

Finally, we divide through by C so that it takes the same form as equation (11.1)

$$\frac{d^2v(t)}{dt^2} + \frac{1}{RC}\frac{dv(t)}{dt} + \frac{1}{LC}v(t) = 0$$

Determine the response type To determine the type of response (under, over, critically damped), check the value of $b^2 - 4c$

$$\left(\frac{1}{RC}\right)^2 - \frac{4}{LC} = \left(\frac{1}{20\,\Omega \times 40\,\mu\text{F}}\right)^2 - \frac{4}{10\,\mu\text{H} \times 40\,\mu\text{F}} = -9998437500 < 0$$

The value is negative, so this is the underdamped case, and the response follows equation (11.9).

Underdamped parameters From our treatment of the underdamped system, we know that

$$\begin{aligned}
\sigma &= \frac{b}{2} = \frac{1}{2RC} = 625 \\
\omega_d &= \frac{\sqrt{4c - b^2}}{2} = \frac{\sqrt{\frac{4}{LC} - \left(\frac{1}{RC}\right)^2}}{2} = 49996.1 \\
v(0) &= 10\,\text{V} = A + \frac{F}{c} = A + \frac{0}{1/LC} \implies A = 10 \\
v'(0) &= \frac{i_c(0)}{C} = 0\,\text{A} = B\omega_d - A\sigma = 49996.1B - 10 \times 625 \implies B = 0.125
\end{aligned}$$

The response Putting it all together, the expected response is

$$\begin{aligned}
v(t) &= e^{-625t}(10\cos 49996.1t + 0.125\sin 49996.1t)\,\text{V} \\
&= 10e^{-625t}\cos(49996.1t + 0.1249)\,\text{V}
\end{aligned}$$

In other words, the voltage response is a decaying sinusoid with frequency 7957.1 Hz and 7.2° phase shift, as seen in figure 11.6.

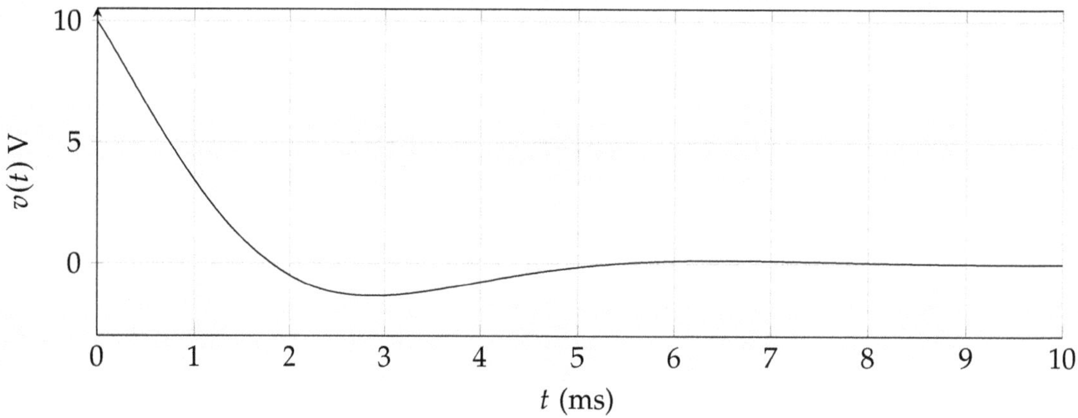

Figure 11.6: Predicted voltage response of the circuit shown in figure 11.5 with $v_c(0) = 0\,\text{V}$ and $i_c(0) = 0\,\text{A}$.

Circuit currents We can compute the circuit currents from $v(t)$.

The resistor's current is simply

$$i_{\text{R}}(t) = \frac{v(t)}{R} = \frac{1}{2}e^{-625t}\cos(49996.1t + 0.1249)\,\text{A}$$

The capacitor's current can be found with the capacitor equation

$$\begin{aligned} i_C(t) &= C\frac{dv(t)}{dt} \\ &= -6250e^{-625t}(\cos(49996.1t + 0.1249) + 80\sin(49996.t + 0.1249)) \\ &= -500000e^{-625t}\sin(49996.1t + 0.1374)\text{ A} \end{aligned}$$

Finally, the inductor's current can be found using Kirchhoff's current law

$$\begin{aligned} i_L(t) &= -i_R(t) - i_C(t) = \\ &= -\frac{1}{2}e^{-625t}\cos(49996.1t + 0.1249) + 500000e^{-625t}\sin(49996.1t + 0.1374) \end{aligned}$$

Example 11.3.2: Solving a series RLC circuit in an overdamped response

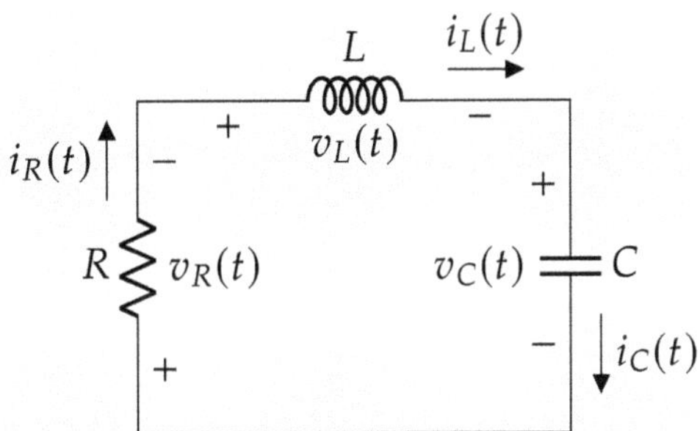

Figure 11.7: A series RLC circuit.

Determine the voltage, $v_C(t)$. Let $L - 20\,\mu\text{H}$, $C - 10\,\mu\text{F}$, and $R - 100\,\Omega$. The initial conditions are $i_C(0) = 0.25$ A and $v_L(0) = 0$ V.

Find the system equation We begin by finding the system equation. Notice that all of the components in figure 11.7 are in series, therefore

$$i(t) = i_R(t) = i_L(t) = i_C(t)$$

Using Kirchhoff's voltage law, we have

$$\begin{aligned} v_R(t) + v_L(t) + v_C(t) &= 0 \\ Ri(t) + L\frac{di(t)}{dt} + \frac{1}{C}\int i(t)\,dt &= 0 \end{aligned}$$

Now we take the derivative of both sides to eliminate the integral.

$$L\frac{d^2i(t)}{dt^2} + R\frac{di(t)}{dt} + \frac{1}{C}i(t) = 0$$

Finally, we divide through by L so that it takes the same form as equation (11.1)

$$\frac{\mathrm{d}^2 i(t)}{\mathrm{d}t^2} + \frac{R}{L}\frac{\mathrm{d}i(t)}{\mathrm{d}t} + \frac{1}{LC}i(t) = 0$$

Determine the response type To determine the type of response (under, over, critically damped), check the value of $b^2 - 4c$

$$\left(\frac{R}{L}\right)^2 - \frac{4}{LC} = \left(\frac{100\,\Omega}{20\,\mu\mathrm{H}}\right)^2 - \frac{4}{20\,\mu\mathrm{H} \times 10\,\mu\mathrm{F}} = 2.498 \times 10^{13} > 0$$

The value is positive, so this is the overdamped case and the response follows equation (11.7).

Overdamped parameters From our treatment of the overdamped system, we know that

$$s_1 = \frac{-b + \sqrt{b^2 - 4c}}{2} = \frac{-\frac{R}{L} + \sqrt{\left(\frac{R}{L}\right)^2 - \frac{4}{LC}}}{2} = \frac{-\frac{100}{20\,\mu\mathrm{H}} + \sqrt{\left(\frac{100\,\Omega}{20\,\mu\mathrm{H}}\right)^2 - \frac{4}{20\,\mu\mathrm{H}\times 10\,\mu\mathrm{F}}}}{2} = -1000.2$$

$$s_2 = \frac{-b - \sqrt{b^2 - 4c}}{2} = \frac{-\frac{R}{L} - \sqrt{\left(\frac{R}{L}\right)^2 - \frac{4}{LC}}}{2} = \frac{-\frac{100}{20\,\mu\mathrm{H}} - \sqrt{\left(\frac{100\,\Omega}{20\,\mu\mathrm{H}}\right)^2 - \frac{4}{20\,\mu\mathrm{H}\times 10\,\mu\mathrm{F}}}}{2} = -4.99 \times 10^6$$

$$i(0) = 0.25\,\mathrm{A} = K_1 + K_2 + \frac{F}{1/LC} = K_1 + K_2 + \frac{0}{1/LC} \implies K_1 + K_2 = 0.25\,\mathrm{A}$$

$$i'(0) = \frac{v(0)}{L} = 0\,\mathrm{V} = s_1 K_1 + s_2 K_2 \implies -1000.2K_1 - 4.99 \times 10^6 K_2 = 0\,\mathrm{V}$$

Solving the last two conditions simultaneously, we have

$$K_1 = 0.25 \qquad\qquad K_2 = -0.00005$$

The current response Putting it all together, the expected response for the current is

$$i(t) = 0.25e^{-1000.2t} - 0.00005e^{-4.99\times 10^6 t}\ \mathrm{A}$$

The circuit voltages We can compute the circuit voltages from $i(t)$.

The resistor's voltage is simply

$$v_\mathrm{R}(t) = i_\mathrm{R}(t)R = 25e^{-1000.2t} - 0.005e^{-4.99\times 10^6 t}\ \mathrm{V}$$

The inductor's voltage can be found with the inductor equation

$$v_\mathrm{L}(t) = L\frac{\mathrm{d}i(t)}{\mathrm{d}t} = -0.5001e^{-1000.2t} + 0.499e^{-4.99\times 10^6 t}\ \mathrm{V}$$

Finally, the capacitor voltage can be found using Kirchhoff's voltage law

$$v_C(t) = -v_R(t) - v_L(t) = \\ = 24.5e^{-1000.2t} - 0.494e^{-4.99\times10^6 t}\ \text{V}$$

In other words, the voltage response is two decaying exponentials, as seen in figure 11.8.

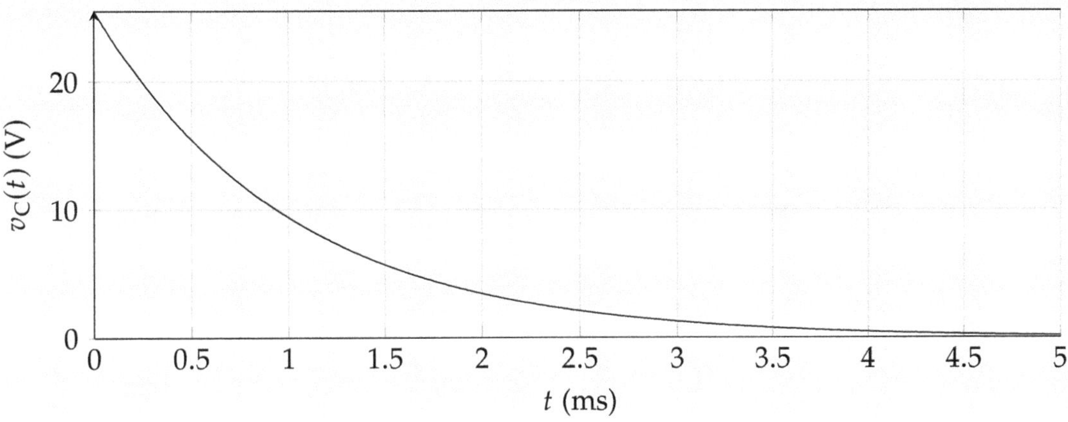

Figure 11.8: Predicted voltage response of the circuit shown in figure 11.7 with $i_C(0) = 0.25\,\text{A}$ and $v_L(0) = 0\,\text{V}$.

11.3.1 Series and parallel RLC response summary

All second order RLC circuits will either follow the parallel or series response even if the circuit does not at first appear to be parallel or series. In order to identify if a circuit is parallel or series, observe the connection between the inductor and capacitor. Whether or not the capacitor and inductor are in series or parallel defines the operation of the circuit. When in doubt, you can always use Kirchhoff's voltage law (KVL) or Kirchhoff's current law (KCL) to write the differential equation for the circuit and compare it to the series and parallel equations. Additionally, source transforms and Thévenin equivalent circuits can be used to simplify the sources and resistors so that it looks like one of the ideal cases in figure 11.9.

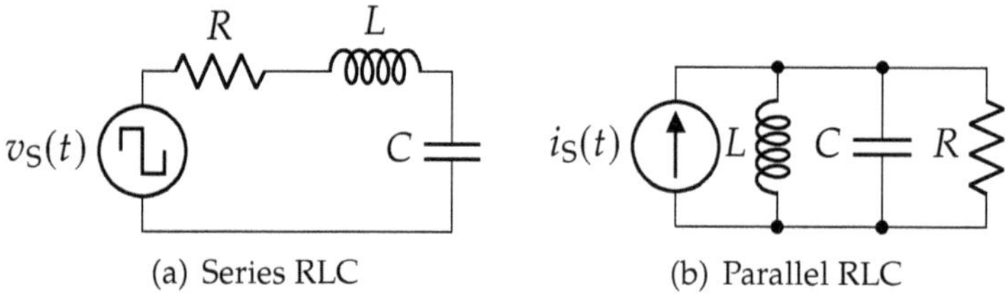

(a) Series RLC

(b) Parallel RLC

Figure 11.9: Simplest series and parallel circuits

The characteristic equation for an RLC circuit takes one of two forms:

$$s^2 + \frac{R}{L}s + \frac{1}{LC} = 0 \quad \text{(Series)}$$
$$s^2 + \frac{1}{RC}s + \frac{1}{LC} = 0 \quad \text{(Parallel)}$$

The roots of the characteristic equation are:

$$s_1, s_2 = \frac{R}{2L} \pm \frac{1}{2}\sqrt{\left(\frac{R}{L}\right)^2 - \frac{4}{LC}} \quad \text{(Series)} \qquad (11.10)$$

$$s_1, s_2 = \frac{1}{2RC} \pm \frac{1}{2}\sqrt{\left(\frac{1}{RC}\right)^2 - \frac{4}{LC}} \quad \text{(Parallel)} \qquad (11.11)$$

All voltages and currents in a series or parallel circuit will have one of the following three forms depending on the solution to the characteristic equation.

$$x(t) = \begin{cases} K_1 e^{s_1 t} + K_2 e^{s_2 t} + x(\infty) & s_1, s_2 \in \mathbb{R} \\ (K_1 + tK_2)e^{st} + x(\infty) & s = s_1 = s_2 \in \mathbb{R} \\ e^{-\sigma t}(A\cos(\omega_d t) + B\sin(\omega_d t)) + x(\infty) & s_1, s_2 \in \mathbb{C} \end{cases}$$

As a reminder, these responses are called overdamped, critically damped, and underdamped, respectively.

The overdamped coefficients K_1 and K_2 are calculated by using the initial conditions.

$$x(0) = K_1 + K_2 + x(infty)$$
$$x'(0) = s_1 K_1 + s_2 K_2$$

In the critically damped case, the coefficients are:

$$x(0) = K_1 + x(\infty)$$
$$x'(0) = sK_1 + K_2$$

Recall that inductors have continuous current and capacitors have continuous voltage. That is, $i_L(t^-)i = i_L(t^+)$ and $v_C(t^-) = v_C(t^+)$.

The initial conditions for $x'(0)$ can be found by using the characteristic equations for the inductor and capacitor. This yields $v_C'(0) = \frac{1}{C}i_C(0^+)$ for a capacitor and $i_L'(0) = \frac{1}{L}v_L(0^+)$. Note that both of these are the current or voltage evaluated at time $t = 0^+$.

For the underdamped case, the roots have the form $s_1, s_2 = -\sigma + j\omega_d$, so the simplified relations are

$$\sigma = \frac{R}{2L} \qquad \omega_d = \sqrt{\frac{1}{LC} - \left(\frac{R}{2L}\right)^2} \qquad \text{(Series)}$$

$$\sigma = \frac{1}{2RC} \qquad \omega_d = \sqrt{\frac{1}{LC} - \left(\frac{1}{2RC}\right)^2} \qquad \text{(Parallel)}$$

ω_d is called the "damped" oscillation frequency while $\omega_0 = \frac{1}{\sqrt{LC}}$ is the "natural" oscillation frequency. ω_d and ω_0 are related by $\omega_d^2 = \omega_0^2 - \sigma^2$.

The coefficients A and B are found using the initial conditions:

$$x(0) = A + x(\infty)$$
$$x'(0) = -A\sigma + B\omega_d$$

Each solution contains an $x(\infty)$ term. This term represents the "final" value that the circuit will reach given unlimited time. The final value can be found by replacing inductors with short circuits and capacitors with open circuits then solving for the desired voltage or current.

11.4 Prelab

Task 11.4.1: Prelab Questions

1. *Calculate* the step response $v_C(t)$ for $t > 0$ for the circuit in figure 11.10.

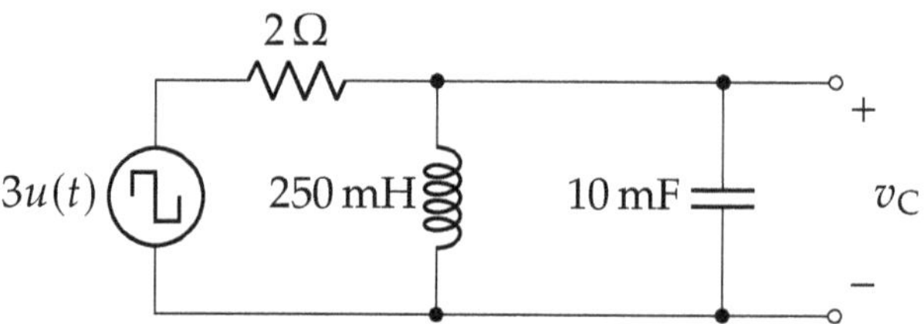

Figure 11.10: Parallel RLC Circuit

2. *Calculate* the Thévenin equivalent resistance shown in figure 11.11 by solving for the ratio v_{test}/i_{test}.

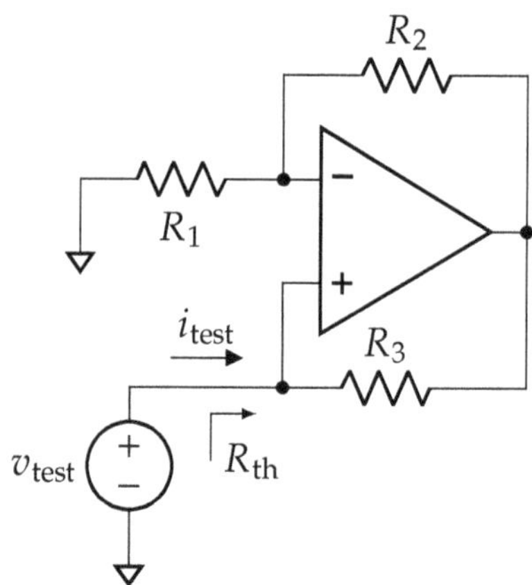

Figure 11.11: A negative impedance converter circuit

3. *Plot* the first 2 ms of $v_C(t)$ in figure 11.12 if $v_C(0) = 1\,\text{mV}$ and $i_L(0) = 0$ using Python's NumPy, MATLAB®, or a similar plotting tool.

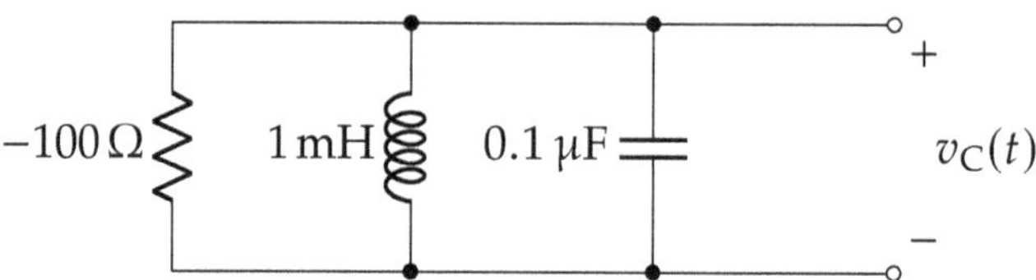

Figure 11.12: A parallel RLC circuit

11.5 Tasks

Task 11.5.1: Series RLC response measurements

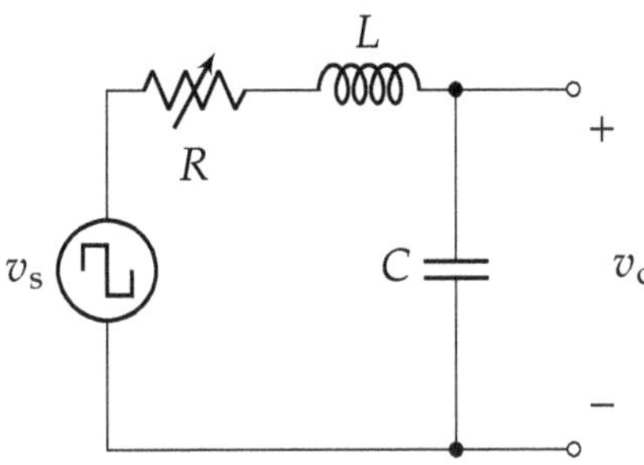

Figure 11.13: A series RLC circuit.

For this task, use $L = 1\,\text{mH}$ and $C = 0.001\,\mu\text{F}$. Assume zero initial conditions. Even though the input square wave *is* a time varying signal, if the period is much longer than the response, then it is effectively constant. One half of the square wave sets the initial conditions, and the other half provides a constant input.

1. Let v_s be a square wave that varies between 0 V and 3 V at 1 kHz. *Adjust* the potentiometer R to obtain and *capture* an underdamped and overdamped v_c response.
2. *Calculate* the peak value of v_s and *compute* a resistance value[a] to obtain the response

$$v_C(t) = 3.5e^{-1.558\times10^6 t} - 8.5e^{-642\times10^3 t} + 4$$

3. *Replace* the potentiometer with the fixed resistor computed in the last step. *Capture* the waveform v_c by *applying* the correct input and *downloading* the data from the oscilloscope.
4. *Use* Python's NumPy or MATLAB®, to *plot* and *compare* data collected from the oscilloscope with the expected response.

[a] Hint: compute $s_1 + s_2$ using equation (11.10).

Task 11.5.2: Parallel RLC response measurements

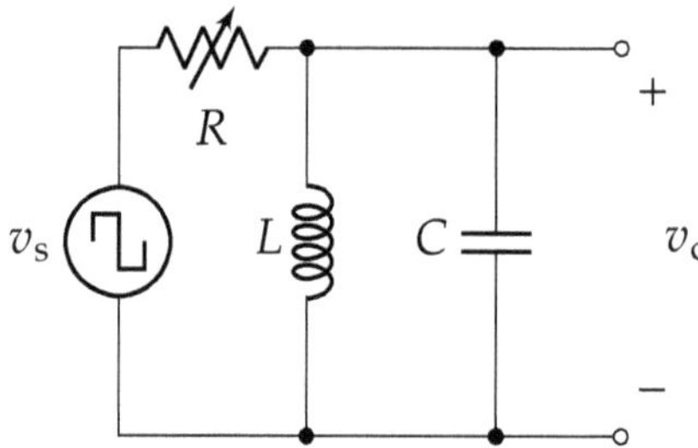

Figure 11.14: A parallel RLC circuit.

For this task, use $L = 1\,\text{mH}$ and $C = 2\,\text{nF}$. Assume zero initial conditions. Even though the input square wave *is* a time varying signal, if the period is much longer than the response, then it is approximately constant. One half of the square wave sets the initial conditions and the other half provides a constant input.

1. Let v_s be a square wave that varies between 0 V and 3 V at 1 kHz. *Adjust* the potentiometer R to obtain and *capture* an underdamped and overdamped v_c response.

2. *Determine* the peak value of v_s and *compute* a resistance value to obtain the response

$$v_C(t) = 1.3e^{-9.2\times10^4 t}(\sin 7 \times 10^5 t)$$

3. *Replace* the potentiometer with the fixed resistor computed in the last step. *Capture* the waveform v_c by applying the correct input and *downloading* the data from the scope.

4. *Use* Python's NumPy or MATLAB® to *plot* and *compare* data collected from the oscilloscope with the expected response.

Task 11.5.3: Oscillator

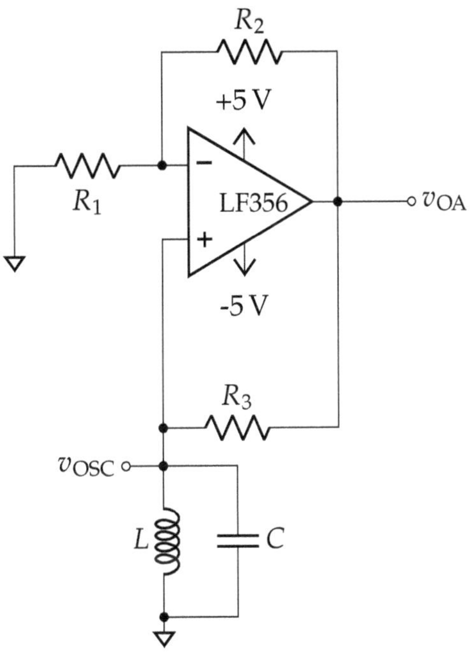

Figure 11.15: An oscillator circuit.

For this task, use $R_1 = R_2 = 1\,\text{k}\Omega$, $R_3 = 100\,\Omega$, $L = 1\,\text{mH}$, and $C = 0.1\,\mu\text{F}$.

1. *Measure* the ESR of the inductor.
2. *Simulate* the circuit in figure 11.15. Use the measured ESR values for the inductor. *Obtain* plots of v_{OA} and v_{OSC}.
3. *Construct* the oscillator circuit.
4. *Capture* an oscilloscope screenshot showing v_{OUT} and v_{OSC} with frequency and peak to peak measurements.
5. *Calculate* the error between the simulated frequency and measured frequency. How do these relate to ω_d for the circuit?

EXPERIMENT 12

Switching circuits: DC voltage regulation

12.1 Application

The simple RLC circuit is used in almost every power supply and electronic device sold today. With the addition of a couple of switches, an RLC circuit can be used to efficiently regulate voltages to different levels. While the circuits in actual devices contain more sophisticated control logic, the core method for changing voltage levels is the RLC circuit.

Switching voltage converters are used to allow ultra lower power devices get usable voltages to run their circuitry as well as step high voltage and high power supplies down to low voltages very efficiently. They are essential to any modern circuit or system design.

Figure 12.1: An array of switching voltage regulators on a motherboard. Multiple regulators are necessary as some processors can continuously draw over 100 A!

12.2 Electronic switches

The original electronic switch was the electromechanical relay. It involved using an electromagnet to open and close a physical metal switch. They are still used today to turn high power loads on and off, but they cannot switch fast enough for the RLC circuits we are building.

Instead of relays, we will be using "solid-state" switches. There are several types of semiconductor devices that can be used as electronically controlled solid state switches. The actual physics of their operation is not needed to use these switches in circuits. Instead, we will assume that the devices are short circuits while turned on and open circuits when turned off.

12.2.1 Metal-oxide-semiconductor field-effect transistors (MOSFETs)

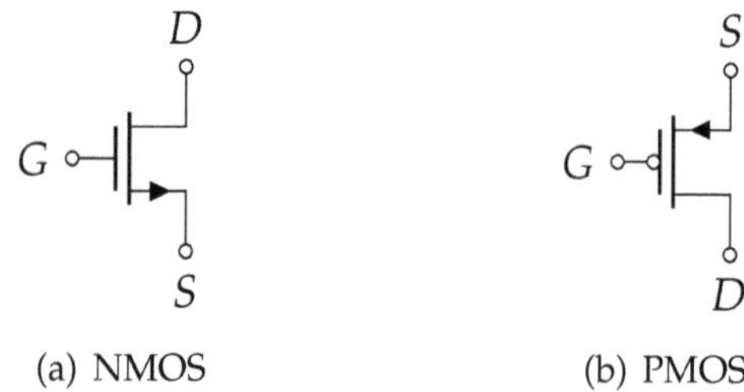

Figure 12.2: MOSFET symbols

The MOSFET is a three terminal device represented by the symbols in figure 12.2. The switch model for the N channel MOSFET is in figure 12.3(a) and the P channel MOSFET is in figure 12.3(b).

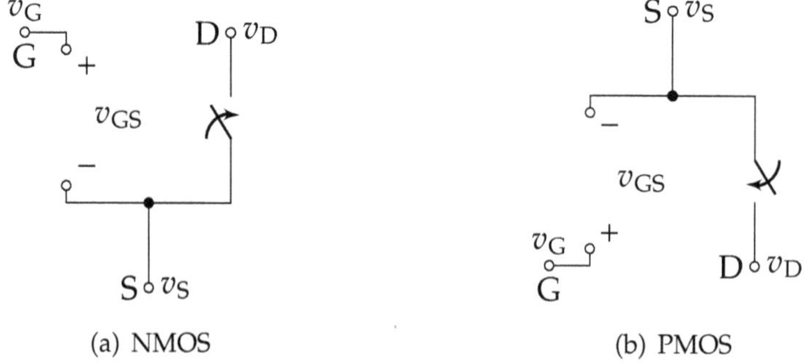

Figure 12.3: MOSFET switch models

For the N channel MOSFET, if $V_G - V_S \gg V_{th}$[1], then the connection between the D and S terminals becomes an effective

[1] When V_{GS} is close to V_{th}, the MOSFET does not work properly as a switch. V_{GS} should be a few volts higher than the threshold voltage to make sure the switch turns on completely.

short circuit. Otherwise, the connection between the D and S terminals is an open circuit. The threshold voltage V_{th} varies by device, but is most commonly from 1 V to 3 V.

The P channel MOSFET acts like the opposite of the N channel MOSFET. The connection between the S and D terminals becomes a short circuit when $V_S - V_G \gg V_{th}$. Both N and P channel MOSFETs can be used to control switching in RLC circuits.

12.2.2 Diodes

The diode is a two-terminal device represented by the symbol in figure 12.4(a). It only allows current to flow in one direction, much like a check valve. The ideal diode acts like a short circuit when $V_A > V_C$ and an open circuit when $V_A < V_C$. The switch model is shown in figure 12.4(b).

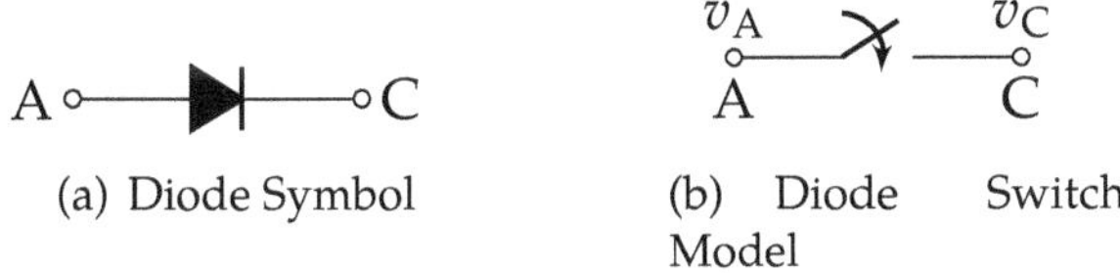

(a) Diode Symbol (b) Diode Switch Model

Figure 12.4: The Diode

12.3 Switched RLC circuits

In this experiment, we will analyze three different switched RLC circuits. Each circuit uses two switches in order to control the circuit behavior. We will be using a function generator (FG) to control the MOSFET switches, but in actual designs there are dedicated integrated circuits (ICs) that would generate the correct frequency square waves to control the switches.

12.3.1 Buck converter

The buck converter is an application of an RLC circuit that takes a large input voltage and efficiently produces a smaller output voltage. The circuit for a buck converter is shown in figure 12.5. The buck converter has three stages:

1. S_1 is closed and S_2 is open. The circuit is a driven RLC circuit.

2. S_1 is open and S_2 is closed. The circuit is an undriven RLC circuit.

3. Both switches are open. The circuit is a first order RC circuit.

The duration of the first stage is set by a control signal. The control signal is a square wave with period T and duty cycle D. The second stage allows the inductor to fully discharge into the circuit. As soon as $i_L(t) = 0$, the circuit will move into the third stage with both switches open. Depending on the values of the components and switching frequency, the circuit may never enter the third stage.

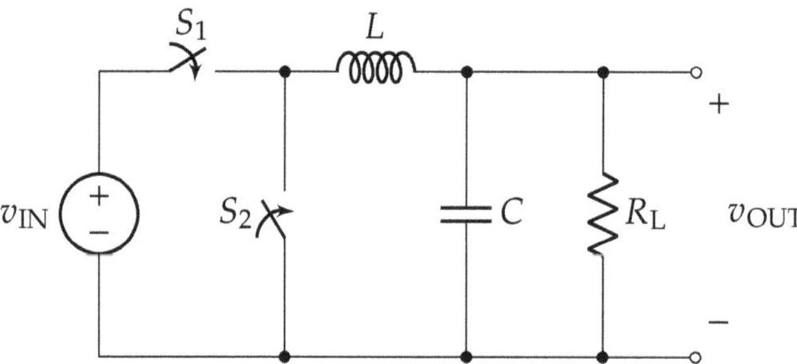

Figure 12.5: A buck converter circuit.

In order to understand how this circuit works, we must find the equilibrium response. The circuit is at equilibrium when the voltage and current at the beginning and the end of a period are equal. As long as the load stays the same, every period will have the same behavior. We can apply the equilibrium conditions to find equation (12.1) and equation (12.2).

$$\begin{aligned} i_L(T) &= i_L(0) \\ i_L(T) - i_L(0) &= 0 \\ \frac{1}{L}\int_0^T v_L(t)\,dt &= 0 \end{aligned} \tag{12.1}$$

$$\begin{aligned} v_C(T) &= v_C(0) \\ v_C(T) - v_C(0) &= 0 \\ \frac{1}{C}\int_0^T i_C(t)\,dt &= 0 \end{aligned} \tag{12.2}$$

A full analysis of the RLC circuits can be performed to predict the response of the converter; however, finding the equilibrium point without any assumptions is difficult.[(2)] The key assumption needed is that the output voltage has negligible ripple. This assumption can be made as long as the time constant for the circuit $\tau = RC$ is much larger than the switching period T. This typically involves picking a sufficiently large capacitor or increasing the switching frequency. With this assumption, we can treat the output as a DC voltage.

(2) As part of the design process for a converter, it is useful to make a first order estimate with assumptions then simulate the response using SPICE.

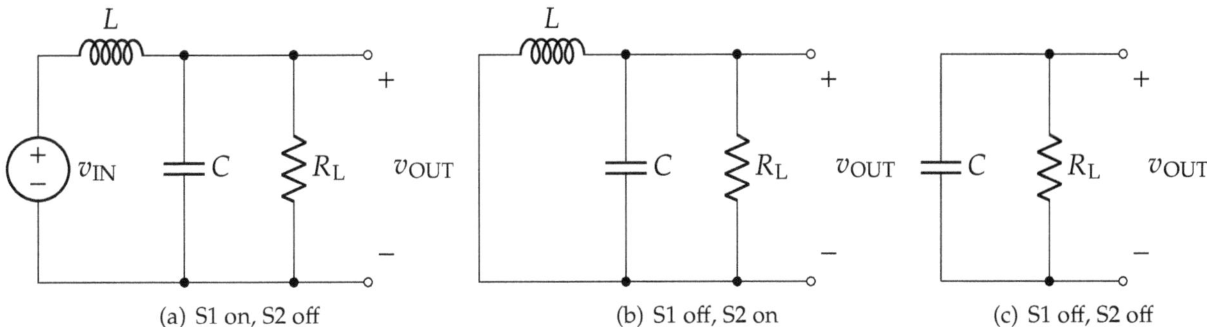

Figure 12.6: Buck converter operation

Finding $i_L(t)$ In figure 12.6(a), we can use the assumption that the output voltage is effectively constant to say that $v_L(t) = v_{IN} - v_{OUT}$. After DT seconds pass, S_1 opens and S_2 closes, yielding the circuit in figure 12.6(b). In this stage, $v_L(t) = -v_{OUT}$. Once the current in the inductor drops to 0, S_2 opens, and the circuit becomes figure 12.6(c). In this stage, $v_L(t) = 0$. When all three stages occur, $i_L(T) = i_L(0) = 0$, so the equilibrium condition is satisfied.

These can be plugged into equation (12.1) to find the inductor current.

$$i_L(t) = \begin{cases} \frac{1}{L}\int_0^t v_{IN} - v_{OUT}\,ds = \frac{t}{L}(v_{IN} - v_{OUT}) & 0 < t \le DT \\ i_L(DT) + \frac{1}{L}\int_{DT}^t -v_{OUT}\,ds = \frac{DT}{L}(v_{IN} - v_{OUT}) - \frac{t-DT}{L}v_{OUT} & DT < t \le DT\frac{v_{IN}}{v_{OUT}} \\ 0 & DT\frac{v_{IN}}{v_{OUT}} < t \le T \end{cases}$$

A plot of the inductor current is shown in figure 12.7

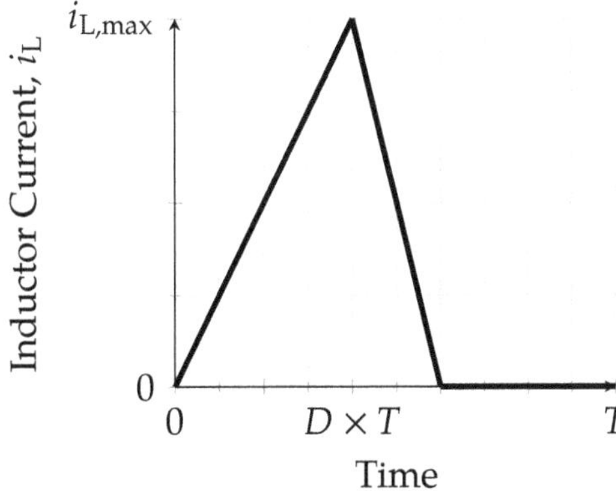

Figure 12.7: Buck converter inductor current

Current is a measure of coulombs per second, so the average current over a period can be used to measure the total amount of charge that enters a circuit. When the circuit is in equilibrium, the same amount of charge must enter in a single period that leaves in a single period; therefore, the average current in needs to be the same as the average current out.

Because the circuit is in equilibrium, all of the energy that enters the circuit must be dissipated by the load in the same period. The net result of this is that the average current through the inductor must be the same as the average current through the load.

$$\frac{1}{T}\int_0^T i_L(t)\,\mathrm{d}t = \frac{v_{\mathrm{OUT}}}{R}$$
$$\frac{1}{2T}DT\frac{v_{\mathrm{IN}}}{v_{\mathrm{OUT}}}\frac{DT}{L}(v_{\mathrm{IN}} - v_{\mathrm{OUT}}) = \frac{v_{\mathrm{OUT}}}{R}$$
$$v_{\mathrm{OUT}} = Dv_{\mathrm{IN}}\frac{\sqrt{8RLT + D^2R^2T^2} - DRT}{4L} \tag{12.3}$$

Since $i_L(t)$ is triangular, we can easily evaluate its intergral using $A = \frac{1}{2}BH$, where A is the area of a triangle, B is the length of the base, and H is the height.

If the desired input and output voltages are known, the duty cycle D can be calculated by solving equation (12.3):

$$D = \frac{\sqrt{2v_{\mathrm{OUT}}}}{\sqrt{\frac{RTv_{\mathrm{IN}}}{L}\left(\frac{v_{\mathrm{IN}}}{v_{\mathrm{OUT}}} - 1\right)}} \tag{12.4}$$

Notice that the capacitance does not factor into this equation. That is due to the small ripple assumption. As long as the capacitor is big enough, we can use this solution to solve for the duty cycle needed.

The analysis of the converter is much simpler if the inductor current never reaches zero, as switch S_2 never reopens.

$$i_L(t) = \begin{cases} i_L(0) + \frac{t}{L}(v_{\mathrm{IN}} - v_{\mathrm{OUT}}) & 0 < t \leq DT \\ i_L(0) + \frac{DT}{L}(v_{\mathrm{IN}} - v_{\mathrm{OUT}}) - \frac{t-DT}{L}v_{\mathrm{OUT}} & DT < t \leq T \end{cases}$$

In order to solve for the equilibrium case, we set $i_L(T) = i_L(0)$:

$$i_L(0) = i_L(0) + \frac{DT}{L}(v_{\mathrm{IN}} - v_{\mathrm{OUT}}) - \frac{T - DT}{L}v_{\mathrm{OUT}}$$
$$v_{\mathrm{OUT}} = Dv_{\mathrm{IN}} \tag{12.5}$$

When $D > 1 - \frac{2L}{RT}$, equation (12.5) can be used because the current will never go to 0. Otherwise, equation (12.3) must be used.

12.3.2 Boost converter

Many devices need a larger voltage than supplied by the power source. In order to generate the higher voltages, a boost converter is used. The schematic for a boost converter is shown in figure 12.8

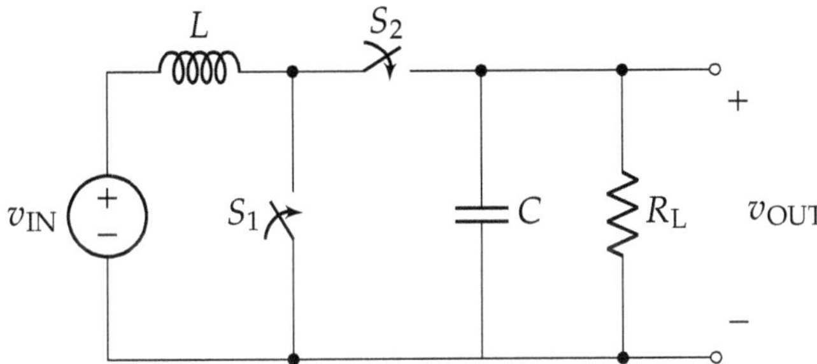

Figure 12.8: A boost converter circuit.

The three stages of operation are:

1. S_1 is closed and S_2 is open. The circuit is an L circuit and an RC circuit.
2. S_1 is open and S_2 is closed. The circuit is a driven RLC circuit.
3. Both switches are open. The circuit is a first order RC circuit.

The time that S_1 stays closed is once again set by the duty cycle D. Then, S_2 remains closed until $i_L(t)$ reaches zero. Just like the buck converter, it is possible that the third state does not occur if the inductor current never reaches zero.

The boost converter is solved using the same procedure as the buck converter. First, the inductor current is found in all three stages.

$$i_L(t) = \begin{cases} \frac{1}{L}\int_0^t v_{\mathrm{IN}}\,\mathrm{d}s = \frac{t}{L}v_{\mathrm{IN}} & 0 < t \le DT \\ i_L(DT) + \frac{1}{L}\int_{DT}^t (v_{\mathrm{IN}} - v_{\mathrm{OUT}})\,\mathrm{d}s = \frac{DT}{L}v_{\mathrm{IN}} + \frac{t-DT}{L}(v_{\mathrm{IN}} - v_{\mathrm{OUT}}) & DT < t \le DT\frac{v_{\mathrm{OUT}}}{v_{\mathrm{OUT}}-v_{\mathrm{IN}}} \\ 0 & DT\frac{v_{\mathrm{OUT}}}{v_{\mathrm{OUT}}-v_{\mathrm{IN}}} < t \le T \end{cases}$$

By setting the average inductor current equal to the average load current, the relationship for input and output can be determined.

$$\frac{D^2 T v_{\mathrm{IN}} v_{\mathrm{OUT}}}{2L(v_{\mathrm{OUT}} - v_{\mathrm{IN}})} = \frac{v_{\mathrm{OUT}}}{R_{\mathrm{L}}}$$

$$v_{\mathrm{OUT}} = v_{\mathrm{IN}}\left(1 + \frac{D^2 RT}{2L}\right) \tag{12.6}$$

$$D = \sqrt{\frac{2L(v_{\mathrm{OUT}} - v_{\mathrm{IN}})}{RT v_{\mathrm{IN}}}} \tag{12.7}$$

In the case where the inductor never fully discharges, the equation for v_{OUT} simplifies:

$$v_{\mathrm{OUT}} = \frac{v_{\mathrm{IN}}}{1 - D} \tag{12.8}$$

Equation (12.8) is only valid when $D(1 - D) < \frac{2L}{RT}$.

12.3.3 Inverting boost converter

Some devices need a negative supply voltage generated from only one DC source. In order to generate the negative voltages, an inverting boost converter is used. The schematic for an inverting boost converter is shown in figure 12.9

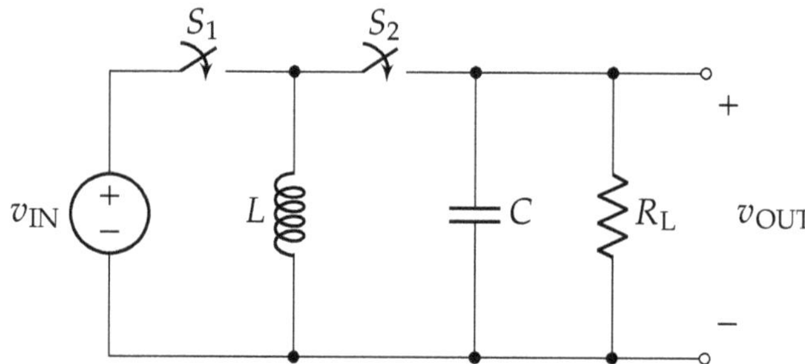

Figure 12.9: An inverting boost converter circuit.

The three stages of operation are:

1. S_1 is closed and S_2 is open. The circuit is an L circuit and a RC circuit.

2. S_1 is open and S_2 is closed. The circuit is an undriven RLC circuit.

3. Both switches are open. The circuit is a first order RC circuit.

We once again start from the inductor current. v_{OUT} will be negative, so keep that in mind when reading the equations.

$$i_L(t) = \begin{cases} \frac{1}{L}\int_0^t v_{\text{IN}}\,ds = \frac{t}{L}v_{\text{IN}} & 0 < t \leq DT \\ i_L(DT) + \frac{1}{L}\int_{DT}^t v_{\text{OUT}}\,ds = \frac{DT}{L}v_{\text{IN}} + \frac{t-DT}{L}v_{\text{OUT}} & DT < t \leq DT\frac{v_{\text{OUT}}-v_{\text{IN}}}{v_{\text{OUT}}} \\ 0 & DT\frac{v_{\text{OUT}}-v_{\text{IN}}}{v_{\text{OUT}}} < t \leq T \end{cases}$$

The relationship between input and output can be determined by setting the average load current equal to the average inductor current.

$$\frac{D^2 T v_{\text{IN}}(v_{\text{OUT}} - v_{\text{IN}})}{2L v_{\text{OUT}}} = -\frac{v_{\text{OUT}}}{R_{\text{L}}}$$

$$v_{\text{OUT}} = -\frac{D v_{\text{IN}}}{4L}\left(RDT + \sqrt{R^2 D^2 T^2 + 8RTL}\right) \quad (12.9)$$

$$D = \sqrt{\frac{2L v_{\text{OUT}}^2}{RT v_{\text{IN}}(v_{\text{IN}} - v_{\text{OUT}})}} \quad (12.10)$$

v_{OUT} again simplifies when $i_L(t)$ never reaches zero:

$$v_{\text{OUT}} = -v_{\text{IN}}\frac{D}{1-D} \quad (12.11)$$

Equation (12.11) is only valid when $D > 1 - \frac{2L}{RT}$; otherwise, the current will reach 0 A and equation (12.9) must be used.

12.4 Prelab

Task 12.4.1: Prelab questions

For the following prelab tasks, use $L = 1\,\text{mH}$ and $C = 10\,\mu\text{F}$. Hint: A resistor can be used as a primitive model of a load with a known voltage and current draw using $R = V_{\text{out}}/I_{\text{load}}$.

1. *Design* the 10 V supply needed in task 12.5.4 by
   a) *Drawing* the overall circuit.
   b) *Selecting* D and T. Use $0.3 < D < 0.7$ and $25\,\text{kHz} < \frac{1}{T} < 100\,\text{kHz}$. You will need to use equation (12.6) or equation (12.8).
2. *Design* the −10 V supply needed in task 12.5.4 by
   a) *Drawing* the overall circuit.
   b) *Selecting* D and T. Use $0.3 < D < 0.7$ and $25\,\text{kHz} < \frac{1}{T} < 100\,\text{kHz}$. You will need to use equation (12.9) or equation (12.11).

Task 12.4.2: Prelab task: Electronic switches

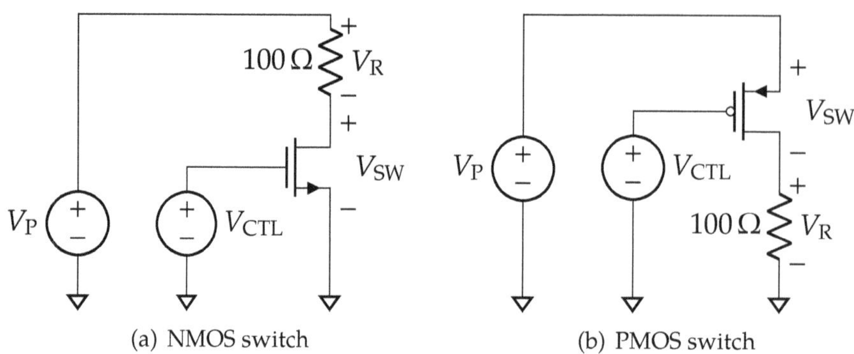

Figure 12.10: NMOS and PMOS test circuits

You may complete this task using the Analog Discovery 2.

1. *Look up* the datasheet for the N-Channel logic level power MOSFET (RFD3055LE) [1] and *find* the pinout for Gate, Drain, and Source. You will use this transistor anywhere an NMOS is used in this lab.
2. *Look up* the datasheet for the P-Channel QFET® power MOSFET (FQU17P06) [2] and *find* the pinout for Gate, Drain, and Source. You will use the transistor anywhere a PMOS is used in this lab.
3. *Construct* the circuit in figure 12.10(a). *Use* a DC supply for V_P and a waveform generator for V_{CTL}.
4. *Set* $V_P = 5\,V$
5. For $V_{CTL} = 0\,V$ and $V_{CTL} = 5\,V$, do the following:
   a) *Measure* V_{SW} and V_R.
   b) *Compute* the current through the switch using $I_{SW} = \frac{V_R}{R}$.
   c) *Compute* the power dissipated in the switch using $P_{SW} = V_{SW} \cdot I_{SW}$.
   d) Is the switch open or closed?
6. *Repeat* step 3 through step 5 using the circuit in figure 12.10(b).

12.5 Tasks

Task 12.5.1: Buck converter

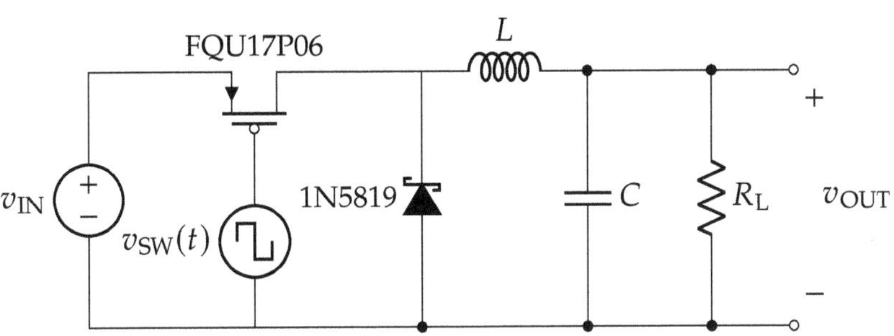

Figure 12.11: A buck converter circuit.

1. *Compute* the theoretical $v_{OUT,avg}$ given $L = 1\,\text{mH}$, $C = 10\,\mu\text{F}$, $R_L = 330\,\Omega$, $v_{IN} = 5\,\text{V}$, and $v_{SW}(t)$ is a $-5\,\text{V}$ to $5\,\text{V}$ 100 kHz square wave with duty cycle $D = 50\%$.
2. *Construct* the circuit in figure 12.11 with the above values.
3. *Measure* the v_{pp} and $v_{OUT,avg}$ with the oscilloscope set to a vertical scale of 100 mV/div. *Calculate* the error for $v_{OUT,avg}$.
4. *Capture* two screenshots of output $v_{OUT}(t)$: One with a vertical scale of 100 mV/div and one with a vertical scale of 1V/div.
5. *Measure* the efficiency of the conversion P_{out}/P_{in}.[a]
6. *Measure* and *plot* $v_{OUT,avg}$ vs. duty cycle of the switch. For this circuit, the duty cycle of the switch is not the duty cycle of the function generator. Instead, $D_{SW} = 1 - D_{FG}$.

[a] Assume that P_{in} can be calculated from the voltage and current reading on the power supply unit. You may omit this step if your power supply unit does not indicate the output current.

Task 12.5.2: Boost converter

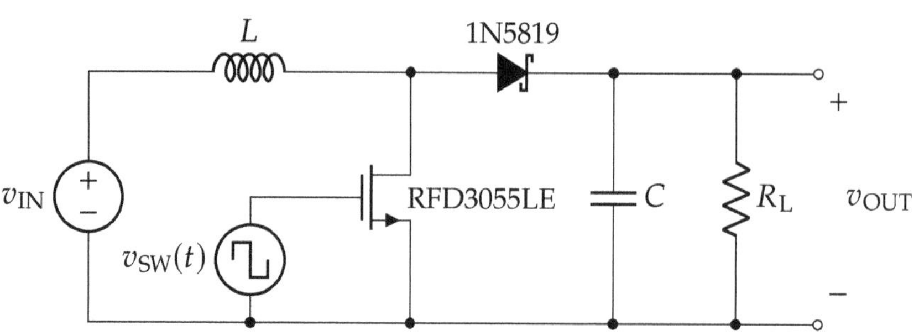

Figure 12.12: A boost converter circuit.

1. *Compute* the duty cycle of the switch needed for $v_{OUT} = 10\,V$ given $L = 1\,mH$, $C = 10\,\mu F$, $R_L = 470\,\Omega$, $v_{IN} = 5\,V$, and $v_{SW}(t)$ is a 0 V to 5 V 75 kHz square wave.
2. *Construct* the circuit in figure 12.12 using the above values.
3. *Measure* the v_{pp} and $v_{OUT,avg}$ with the oscilloscope set to a vertical scale of 200 mV/div. *Calculate* the error for $v_{OUT,avg}$.
4. *Capture* two screenshots of output $v_{OUT}(t)$: One with a vertical scale of 200 mV/div and one with a vertical scale of 1 V/div.
5. *Measure* the efficiency of the conversion P_{out}/P_{in}.[a]

[a] Assume that P_{in} can be calculated from the voltage and current reading on the power supply unit You may omit this step if your power supply unit does not indicate the output current.

Task 12.5.3: Inverted boost converter

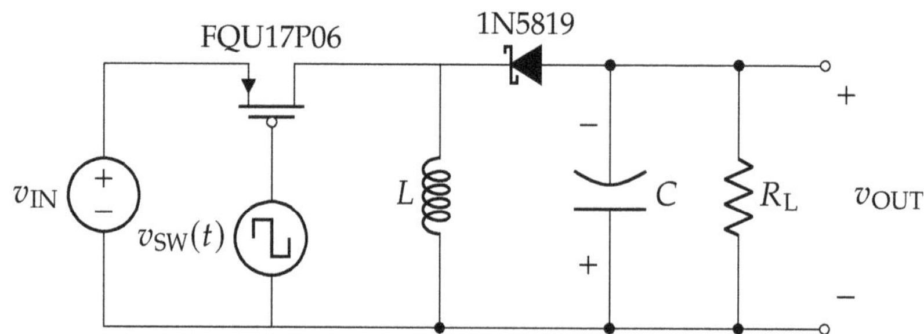

Figure 12.13: An inverted boost converter circuit.

1. *Compute* the duty cycle of the switch needed for $v_{OUT} = -10\,V$ given $L = 1\,mH$, $C = 10\,\mu F$, $R_L = 470\,\Omega$, $v_{IN} = 5\,V$, and $v_{SW}(t)$ is a 0 V to 5 V 100 kHz square wave.
2. *Construct* the circuit in figure 12.13 using the above values. *Recall* that for the PMOS switch, $D_{SW} = 1 - D_{FG}$. Also *ensure* that any electrolytic capacitors are installed in

the correct polarity, as shown in figure 12.13.

3. *Measure* the v_{pp} and $v_{OUT,avg}$ with the oscilloscope set to a vertical scale of 200 mV/div. *Calculate* the error for $v_{OUT,avg}$.

4. *Capture* two screenshots of output $v_{OUT}(t)$: One with a vertical scale of 200 mV/div and one with a vertical scale of 1 V/div.

5. *Measure* the efficiency of the conversion P_{out}/P_{in}.[a]

[a] Assume that P_{in} can be calculated from the voltage and current reading on the power supply unit You may omit this step if your power supply unit does not indicate the output current.

Task 12.5.4: Dual rail power supply design

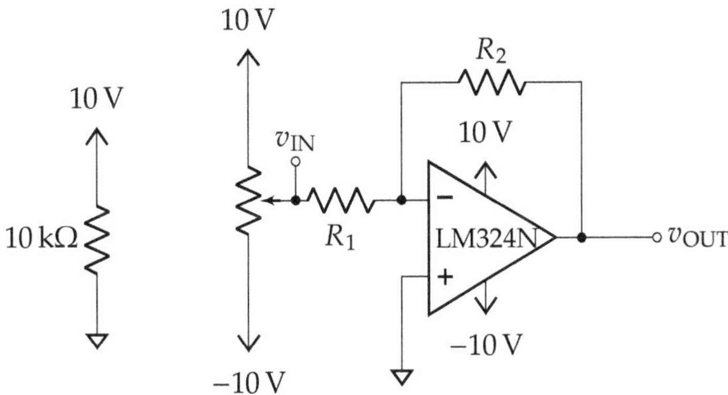

Figure 12.14: An inverting amplifier circuit.

In this task, you will design your own dual rail power supply to drive the operational amplifier in the inverting amplifier circuit from a 5 V USB power supply.

1. *Design* a dual rail power supply that outputs both 10 V and −10 V. The LM324N draws approximately 4 mA. You will need both outputs of the function generator.

2. *Pick* an appropriate R_1 and R_2 so that the op-amp has a gain of −4.7. Pick R_1 and R_2 to both be greater than 10 kΩ.

3. *Construct* the designed circuit with the op amp connected. **Do not turn on the circuit without the opamp connected to the positive supply or the voltage will increase without bound!**

4. *Adjust* the duty cycles[a] on the function generator to fine tune the output voltages to the designed ±10 V.

5. *Measure* v_{IN} and v_{OUT} while varying the potentiometer. *Plot* v_{OUT} vs. v_{IN}.

[a] In a real switching power supply, there are circuits that automatically adjust the duty cycle based on a feedback system.

12.6 References

[1] *RFD3055LE, RFD3055LESM n-channel log level power MOSFET*, RFD3055LE, Rev C0, Fairchild Semiconductor, Sep. 2013. [Online]. Available: https://www.onsemi.com/pub/Collateral/RFD3055LESM-D.pdf.

[2] *FQD17P06/FQU17P06 p-channel QFET® MOSFET*, FQU17P06, Rev C3, Fairchild Semiconductor, Apr. 2014. [Online]. Available: https://www.onsemi.com/pub/Collateral/FQU17P06-D.pdf.

EXPERIMENT 13

Passive Filters

13.1 Application

We have seen in past experiments that circuits may not behave the same under inputs of different frequency. Inductors and capacitors are described with differential equations, and therefore have a frequency-dependent relationship between current and voltage. We leverage this fact to design and build circuits that filter out certain frequencies while letting others through. Such circuits, called filters, are the cornerstone of analog signal processing. While very complex filters have their place, most of the time simple first or second order analog filters are all that is needed.

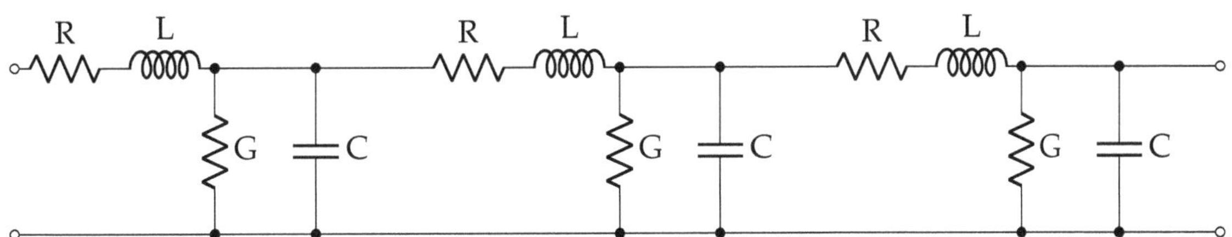

Figure 13.1: RLC model used for a long interconnect or wire. The parasitics model how a cable will filter signals.

13.2 Frequency domain analysis

Mathematical analysis of system frequency response is performed with complex exponentials and phasor analysis. The analysis can be verified in the lab with sinusoidal inputs that sample the response over a range of frequencies. In fact, we have already done this in experiment 7. Every circuit will have some sort of response to signals at different frequencies. This relationship is called the transfer function of the system.

13.2.1 The Fourier series

The Fourier series, shown in equation (13.1), is a way to decompose any periodic signal into a sum of sines and cosines. Because all periodic signals can be written as a Fourier series, we simply need to know the response of a system to each individual frequency in order to determine how it will respond to all periodic signals. The specifics of calculating Fourier series of specific signals is a more advanced topic then we will cover in this experiment; it is only important to understand that knowing a circuit's response to individual sine waves is sufficient to mathematically predict its response to *all* periodic signals.

$$s(t) = \frac{a_0}{2} + \sum_{n=1}^{\infty} a_n \cos(nt) + \sum_{n=1}^{\infty} b_n \sin(nt) \tag{13.1}$$

13.2.2 Complex exponential

For reasons of generality, frequency domain analysis is typically done with complex exponentials. They simplify the notation and allow us to solve for the sine and cosine response at once.(1) To see this, consider Euler's formula

$$e^{j\omega t} = \cos(\omega t) + j\sin(\omega t)$$

where $j = \sqrt{-1}$. From that we have

$$\begin{aligned} \text{Re}\{e^{j\omega t}\} &= \cos(\omega t) \\ \text{Im}\{e^{j\omega t}\} &= \sin(\omega t)\,. \end{aligned} \tag{13.2}$$

That is, the sine and cosine responses are found by simply taking both the real (Re{·}) and imaginary (Im{·}) parts.

All Resistor-Inductor-Capacitor (RLC) circuits can be described using linear differential equations, which, conveniently, implies that a linear circuit with a complex exponential input has a complex exponential output.(2) That is straight from

(1) Actually, that is not the only reason. The Fourier *transform* generalizes these results to non-periodic time domain signals, too. We leave that extension to a more advanced course.

(2) Recall that the "output" of a linear system is simply the voltage or current anywhere in the circuit.

knowing that the solution to a linear differential equation is a complex exponential.(3)Therefore, the response of any RLC circuit must be the same as the input with only potential changes to the magnitude and phase.

(3) You may be thinking that we found solutions to RC and RL circuits that did not involve complex numbers, but those solutions just happen to have complex coefficients and exponentials that cancel out.

Due to the properties shown in equation (13.2), any change to the complex exponentials magnitude and phase results in the same change to the corresponding sine's and cosine's magnitude and phase. That is,

$$\begin{aligned}\mathrm{Re}\{Ae^{j(\omega t+\phi)}\} &= A\cos(\omega t+\phi)\\ \mathrm{Im}\{Ae^{j(\omega t+\phi)}\} &= A\sin(\omega t+\phi)\end{aligned}$$

13.3 Phasors

Another way of stating the above results is that the complex exponential,

$$Ae^{j(\omega t+\phi)},$$

has an analytic, or *phasor*, representation of

$$A\cos(\omega t+\phi).$$

For notational convenience, $A\angle\phi$ is commonly used as short hand. Variables that denote a phasor quantity are bolded. For example, a phasor voltage at node X could be written as

$$\mathbf{V}_{\mathrm{X}} = A\angle\phi$$

Phasors are not always expressed as functions of time. The phasor $Ae^{j\phi}$ may be used to indicate the time varying signal $Ae^{j(\omega t+\phi)}$. With the short hand notation, it is left to a problem's context to determine if the phasor is time varying or not.

13.4 Frequency response

The frequency response of a circuit, $H(w)$, is defined as the magnitude scaling and phase shift of a system over frequency. In other words,

$$H(w) = \frac{v_{\mathrm{OUT}}(\omega)}{v_{\mathrm{IN}}(\omega)}.$$

It turns out that for linear and time invariant (LTI) systems, knowing the transfer function completely specifies the system. While beyond the scope of this lab, it is reassuring to know that the frequency response is not just helpful, but also sufficient. In fact, measuring transfer functions is such a routine task in electrical engineering that there is an entire class of measurement tools to do it called network analyzers.

In the experiment, we can estimate the frequency response by applying sinusoidal inputs of fixed frequency, f, and then measuring the ouput's magnitude and relative phase. We repeat these measurements for all frequencies of interest. A magnitude and phase versus frequency plot is the common method of communicating the data and is called a Bode plot.

13.4.1 Example types of systems

Systems are classified by their general tendencies versus frequency. In general we say a system is low pass, high pass, band pass, or band reject in nature. All of the presented responses are shown on log-log plots. Frequency response plots are almost always shown on semilog or log-log plots due to their exponential nature.(4)

(4) An example of the logarithmic scaling of frequency is musical notes. An A_4 is 440 Hz, $A_5 = 880$ Hz, and $A_6 = 1760$ Hz. The difference between the same notes in separate octaves doubles. If we plot an exponential function on a logarithmic axis, it becomes linear. That makes it easier to analyze visually.

Low pass

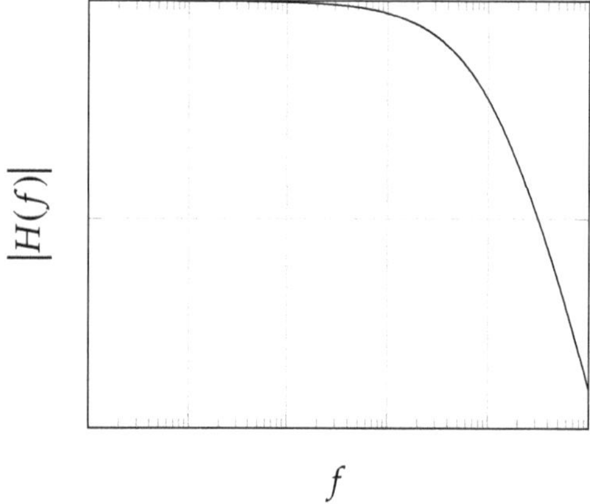

Figure 13.2: Typical frequency response plot of a low pass filter.

Figure 13.2 is a typical low pass filter response. It is called low pass because low frequencies are passed with less loss than high frequencies. Keep in mind that a filter may attenuate *all* frequency content, it just attenuates some more than others. The higher the order(5) filter, the faster the attenuation changes.

(5) Order of differential equation that is needed to describe the circuit.

High pass

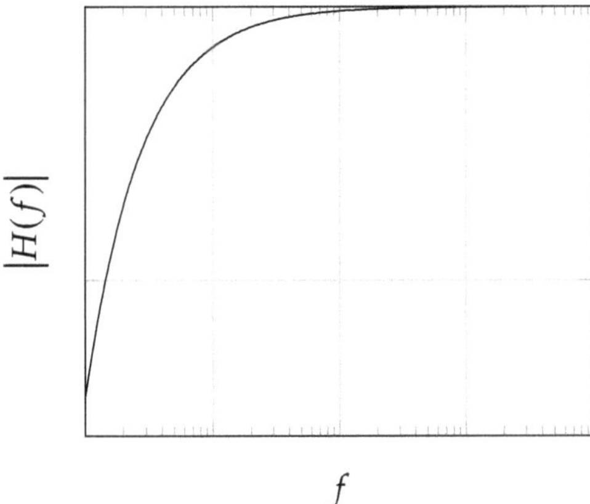

Figure 13.3: Typical frequency response plot of a high pass filter.

Figure 13.2 is a typical high pass filter response. It is called high pass because high frequencies are passed with less loss than low frequencies.

Band pass

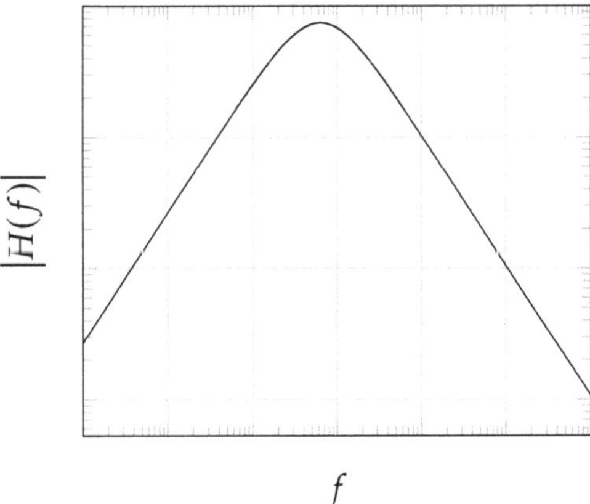

Figure 13.4: Typical frequency response plot of a band pass filter.

Figure 13.4 is a typical band pass filter response. It is called band pass because mid frequencies are passed with less loss than high or low frequencies. Band pass filters can be used to isolate a specific signal in a system.

Band reject (band stop)

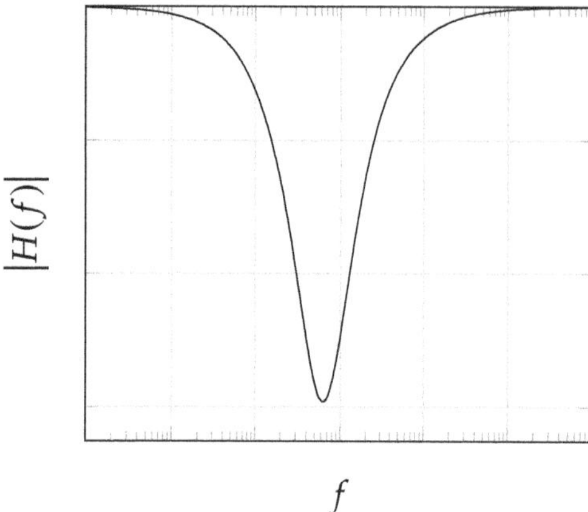

Figure 13.5: Typical frequency response plot of a band reject filter.

Figure 13.5 is a typical band reject filter response. It is called band reject because low and high frequencies are passed with less loss than mid frequencies. Band reject filters can be used to remove an unwanted frequency from a signal.

13.5 Impedance

The impedance of a circuit is a measure of how it resists current flow under an applied voltage. It is defined as the ratio of the complex exponential voltage to the complex exponential current; therefore, it is a complex value. Typically Z is used to denote impedance and has the units Ohms. We have seen the same concept for DC voltages and resistance. In fact, resistance is just the real part of impedance.

Analysis using impedance makes it possible to treat inductors and capacitors as complex resistances and, as a result, dramatically simplies circuit analysis through the generalized form of Ohm's law.

Before we begin, it may be helpful to review basic complex numbers. The two primary forms are polar,

$$Z = Ae^{j\theta},$$

and Cartesian,

$$Z = R + jX$$

where

$$A = |Z| = \sqrt{R^2 + X^2}$$
$$\theta = \angle Z = \arctan\left(\frac{X}{R}\right)$$

R is the impedance's resistance and X is called the reactance. Only the resistive part of impedance consumes power. The reactance represents a storage, and later redelivery, of energy over time.

The inverse tangent function on a calculator typically does not properly account for angles outside of the range $-\frac{\pi}{2} < \theta < \frac{\pi}{2}$; however, most calculators offer a rectangular to polar conversion function that correctly accounts for signs. Additionally, the math library for any programming language should have a function called atan2 that will identify the correct quadrant for the answer.

The polar form is most convenient for multiplying and dividing two impedances, where the coefficients just multiply and divide and the phases add or subtract, respectively. The Cartesian form is best for adding and subtracting, where the real and imaginary parts just add or subtract, respectively.

Definition With that out of the way, we can now define the generalized AC Ohm's law. Let,

$$\mathbf{V} = |\mathbf{V}|e^{j(\omega t+\phi_V)}$$
$$\mathbf{I} = |\mathbf{I}|e^{j(\omega t+\phi_I)}.$$

Then, the complex impedance between the two is

$$Z = \frac{\mathbf{V}}{\mathbf{I}} = \frac{|\mathbf{V}|}{|\mathbf{I}|}e^{j(\phi_V-\phi_I)} \tag{13.3}$$

In other words, the magnitude of the impedance is $|\mathbf{V}|/|\mathbf{I}|$ just as with the DC Ohm's law and the phase lag of the current with respect to the voltage is $\phi_V - \phi_I$.

13.5.1 Impedance of a capacitor

To derive the impdedance of a capacitor, we can directly apply the defination in equation (13.3). We begin with the capacitor equation

$$i_C(t) = C\frac{dv_C(t)}{dt}$$

Let $v_C = Ae^{j\omega t}$ with corresponding phasor $\mathbf{V}_C = A + 0j$. Then,

$$i_C(t) = j\omega CAe^{j\omega t}.$$

$i_C(t)$ has phasor form $\mathbf{I}_C = 0 + j\omega CA$. Thus,

$$Z_C = \frac{\mathbf{V}_C}{\mathbf{I}_C} = \frac{1}{j\omega C} = \frac{1}{j2\pi fC} \tag{13.4}$$

The capacitor is a purely reactive component because the impedance is purely imaginary. An ideal capacitor consumes no real power and produces a current the same shape as the input voltage, but with a 90° phase lead. Small frequencies have a larger impedance than large frequencies, so capacitors tend to "block" low frequency signals by presenting as a large impedance magnitude and "pass" high frequencies like a short.

It is often convenient to cancel imaginary numbers from the denominator of a fraction. For a purely imaginary number, this is done by multiplying by $\frac{j}{j}$: $\frac{1}{j} = \frac{j}{j^2} = -j$. The angle of $-j$ is $-90°$.

13.5.2 Impedance of a inductor

Like the capacitor, to derive the impedance of a inductor, directly apply the definition in equation (13.3). We begin with the inductor equation,

$$v_L(t) = L\frac{\mathrm{d}i_L(t)}{\mathrm{d}t}$$

Let the current $i_L = Ae^{j\omega t}$ with phasor form $\mathbf{I}_L = A + 0j$. Then,

$$v_L(t) = j\omega LAe^{j\omega t}\,.$$

Which has phasor form $\mathbf{V}_L = 0 + j\omega LA$. Thus,

$$Z_L = \frac{\mathbf{V}_L}{\mathbf{I}_L} = j\omega L = j2\pi fL \tag{13.5}$$

The inductor is also a purely reactive component. An ideal inductor consumes no real power and produces a current the same shape as the input voltage, but with a 90° phase lag. Large frequencies have higher impedance than small ones, which makes inductors "pass" low frequency signals by acting like a short and "block" high frequencies by presenting a large impedance magnitude.

13.5.3 Circuit analysis

Most of the circuit properties for resistance generalizes to impedance if one allows everything to become complex.

Series impedance

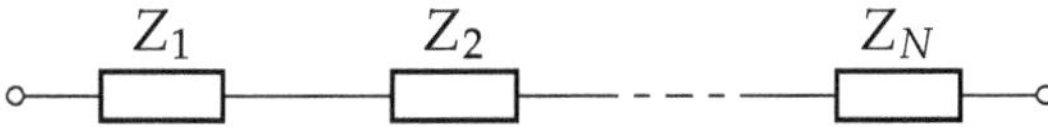

Figure 13.6: Series impedance where Z_n are *any* mix of impedances.

The series impedance formula is the same as with resistances:

$$Z_{\text{total}} = Z_1 + Z_2 + \cdots + Z_N$$

Parallel impedance

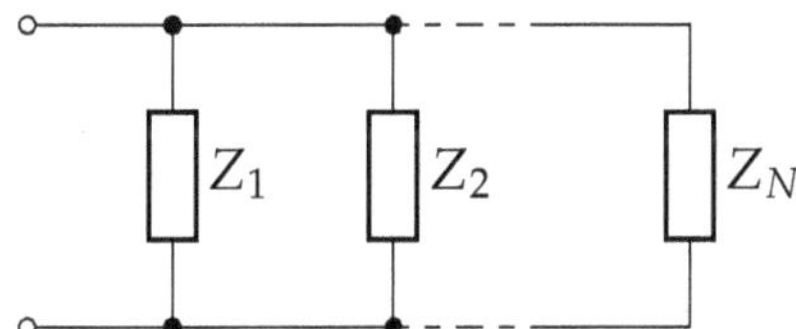

Figure 13.7: Parallel impedance where Z_n are *any* mix of impedances.

The parallel impedance formula is also the same as with resistances:

$$\frac{1}{Z_{\text{total}}} = \frac{1}{Z_1} + \frac{1}{Z_2} + \cdots + \frac{1}{Z_N}$$

When there are only two in parallel the handy simplification is still valid, but the complex arithmetic can lead to mistakes:

$$Z_{\text{total}} = \frac{Z_1 Z_2}{Z_1 + Z_2}$$

Voltage division

While voltage division is just an application of Ohm's law, we state it here for completeness. We only consider the case of two impedances; however, a more complicated circuit can be reduced to this equivalent circuit using the series and parallel combinations techniques from above.

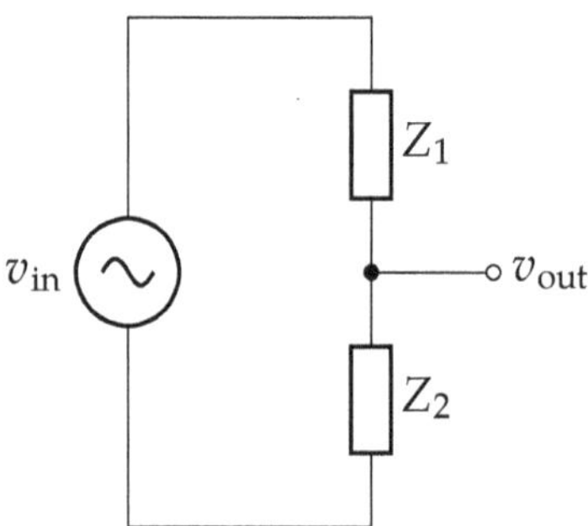

Figure 13.8: Generic voltage division with impedance where Z_n are *any* mix of impedances.

The output can be computed with

$$v_{\text{out}} = \frac{Z_2}{Z_1 + Z_2} v_{\text{in}}$$

The frequency response, or transfer function, is easy to find by simply dividing through by v_{in}.

$$H(f) = \frac{v_{\text{out}}}{v_{\text{in}}} = \frac{Z_2(f)}{Z_1(f) + Z_2(f)}$$

This is the primary method of computing the response for an RLC filter. See example 13.5.1 for a complete example.

Example 13.5.1: Derivation of the transfer function of a low-pass filter

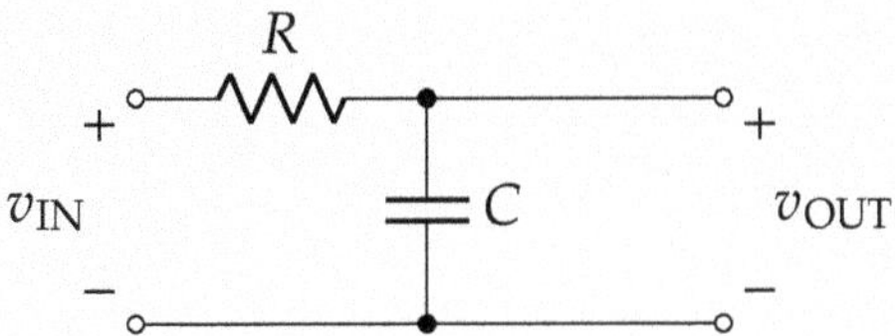

Figure 13.9: An RC filter circuit.

For the RC filter circuit in figure 13.9, we find the circuit element's impedance. Specifically,

$$Z_R = R$$
$$Z_C = \frac{1}{j2\pi fC}$$

Then, to find the transfer function, we simply apply voltage division:

$$H(f) = \frac{v_{OUT}}{v_{IN}} = \frac{Z_C}{Z_C + Z_R} = \frac{\frac{1}{j2\pi fC}}{\frac{1}{j2\pi fC} + R} = \frac{1}{1 + j2\pi fRC}$$

To make more sense out of the complex transfer function, we look at the response magnitude and phase separately.

$$\left|H(f_c)\right| = \frac{1}{\sqrt{1 + (2\pi fRC)^2}}$$
$$\angle H(f) = -\arctan\left(2\pi fRC\right)$$

Clearly the system is low pass. That is, the gain reduces as the frequency increases. Figure 13.10 show an example response of gain (in volts-per-volt) and phase when $R = 1.5\,\text{k}\Omega$ and $C = 0.1\,\mu\text{F}$.

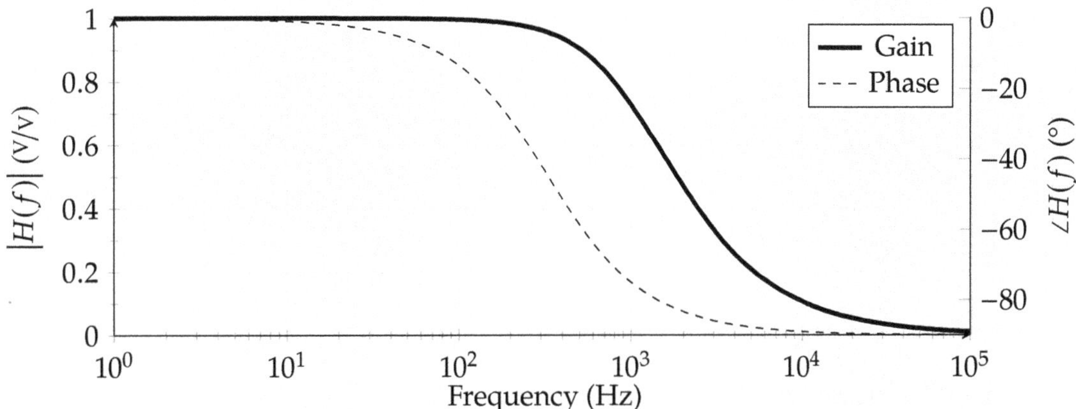

Figure 13.10: Example gain and phase response of a RC low pass filter when $R = 1.5\,\text{k}\Omega$ and $C = 0.1\,\mu\text{F}$ shown as volts-per-volt.

13.6 Decibels

We initially covered the basics of decibels and frequency cutoffs in experiment 10, but we will expand the context to be more general for all filters.

13.6.1 −3 dB Cut-off

The −3 dB point of a system is when the output power is equal to 1/2 the input power. Converted to voltage gain, the half power point is equivalent to:

$$|H(f)| = \frac{1}{\sqrt{2}}$$

13.6.2 Roll-off

The roll off is the decibels of attenuation per one decade of frequency on the blocking side of the cut off point.

13.6.3 −3 dB bandwidth

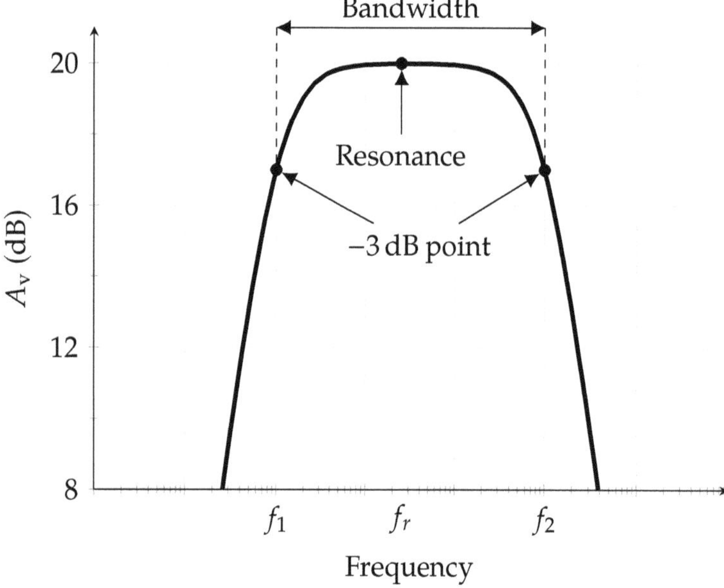

Figure 13.11: Example band pass system with −3 dB and resonance frequency indicated.

The −3 dB bandwidth is the change in frequency between the upper and lower −3 dB points of a band pass/reject system. That is, $f_2 - f_1$ in figure 13.11.

13.6.4 Resonance frequency

The resonance frequency is the peak or minimum of the transfer function in RLC circuits. In general, it occurs when the imaginary part of the impedance cancels out and goes to 0 or infinity. For LC circuits,

$$f_{\mathrm{r}} = \frac{1}{\sqrt{LC}}\ \mathrm{rad} = \frac{1}{2\pi\sqrt{LC}}\ \mathrm{Hz}$$

13.6.5 Q factor

The quality, or Q, factor is a measure of how "peaky" the response is. That is, the ratio of the center, or resonant, frequency to the bandwidth. Specifically,

$$Q = \frac{f_{\mathrm{r}}}{f_2 - f_1}$$

where f_2 is the upper −3 dB point and f_1 is lower.

13.7 Prelab

Task 13.7.1: Prelab questions

1. *Plot* the filter magnitude responses in task 13.7.2 to task 13.8.2 using a tool like MATLAB or Python's NumPy.
2. *Identify* the filter types of the two filters in task 13.8.4.
3. *Record* the input resistance and capacitance of your oscilloscope.

Task 13.7.2: Prelab task: Investigation of an RC filter

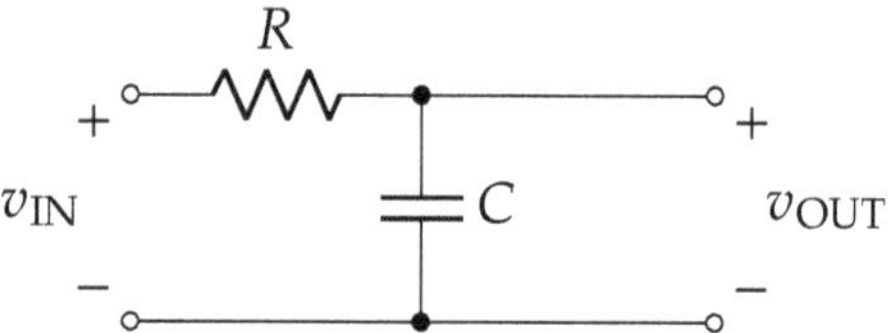

Figure 13.12: An RC filter circuit.

1. *Compute* appropriate R and C values such that the filter's cut-off frequency f_c is 23.873 kHz.
2. *Construct* the circuit in figure 13.12.
3. *Set* v_{IN} to be a 1 $V_{p\text{-}p}$ sine wave with no offset.
4. *Sweep* the frequency of v_{IN} from 10 Hz to 100 kHz. *Record* the magnitude of v_{OUT} and v_{IN} as well as the phase shift from v_{IN} to v_{OUT} at each step. *Record* at least 5 points per decade
5. *Plot* the gain[a] and phase frequency response.
6. *Measure* the −3 dB point by adjusting the input frequency until the output has an amplitude of almost exactly $\frac{1}{\sqrt{2}}$ times the input amplitude. *Capture* an oscilloscope screenshot showing this point with measurements for amplitude and phase shown.

[a] The gain should be in decibels. This is calculated as $G_{db} = 20\log_{10}(G_V(f))$, where $G_V(f)$ is the gain in volts per volt.

13.8 Tasks

Task 13.8.1: Investigation of another RC filter

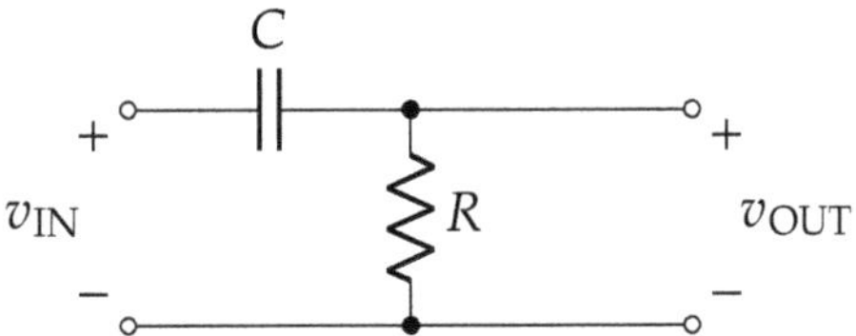

Figure 13.13: Another RC filter circuit.

1. *Compute* appropriate R and C values such that the filter's cut-off frequency f_c is 23.873 kHz.
2. *Construct* the circuit in figure 13.12.
3. *Set* v_{IN} to be a 1 $V_{p\text{-}p}$ sine wave with no offset.
4. *Sweep* the frequency of v_{IN} from 10 Hz to 100 kHz. *Record* the magnitude of v_{OUT} and v_{IN} as well as the phase shift from v_{IN} to v_{OUT} at each step. *Record* at least 5 points per decade
5. *Plot* the gain in decibels and phase frequency response.
6. *Measure* the −3 dB cutoff by adjusting the input frequency until the output has an amplitude of almost exactly $\frac{1}{\sqrt{2}}$ times the input amplitude.
7. *Capture* an oscilloscope screenshot showing this point with measurements for amplitude and phase shown.

Task 13.8.2: Series RLC filter

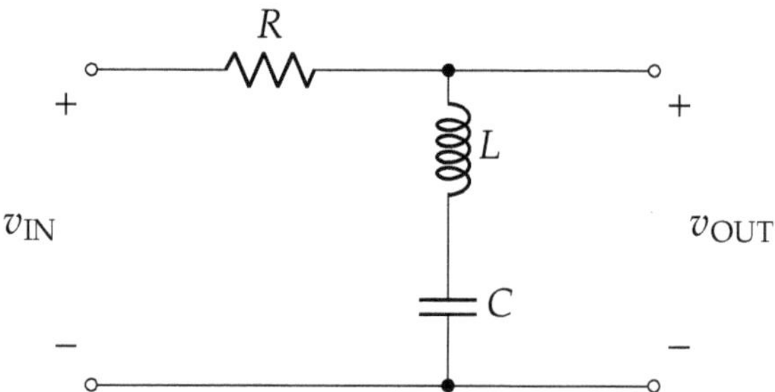

Figure 13.14: An RLC filter circuit.

1. *Compute* the resonance frequency, f_r, for $R = 1\,\mathrm{k}\Omega$, $L = 1\,\mathrm{mH}$ and $C = 1\,\mu\mathrm{F}$.
2. *Construct* the circuit in figure 13.14.
3. *Measure* and *plot* the frequency response (magnitude in decibels and phase) from 10 Hz to 100 kHz. You may have to sample the frequency response more densely near the resonance point.
4. *Measure* the resonance frequency.
5. *Measure* the −3 dB bandwidth.
6. *Compute* the quality factor Q.

Task 13.8.3: Oscilloscope input impedance 

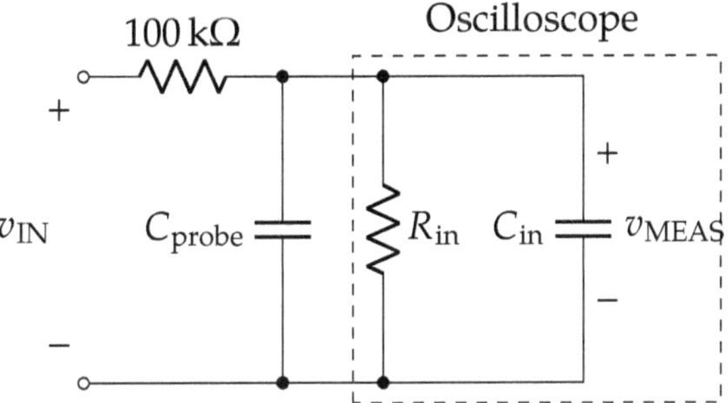

Figure 13.15: Test set for scope input

1. *Calculate* the −3 dB point of the filter created by the oscilloscope input with the probe in x1 mode. Leave C_{probe} as a variable. R_{in} and C_{in} should be printed on the front panel of the oscilloscope.
2. *Construct* the circuit in figure 13.15 using an oscilloscope probe in x1 mode.
3. *Measure* and *plot* the frequency response of the circuit with a probe in x1 mode. *Record* the −3 dB point.
4. *Estimate* the effective cable capacitance from the measured −3 dB point.
5. *Compute* the −3 dB point of the filter with a probe in x10 mode. When in x10 mode, the probe multiplies the **impedance** of the input by 10. Leave C_{probe} as a variable in your equation.
6. *Measure* and *plot* the frequency response of the scope input with a probe in x10 mode. *Record* the −3 dB point.
7. How does the probe mode setting affect the equivalent capacitance C_{probe}?
8. What is the published bandwidth of your scope and scope probe? Do they agree with your measurements? What could be causing any discrepancies between the measured and published bandwidth?

Task 13.8.4: Crossover Analysis

Many loudspeakers contain both a tweeter for high frequency sounds and a woofer for low frequency signals. In order to split the energy between the two speakers, a crossover is needed to filter the signals.

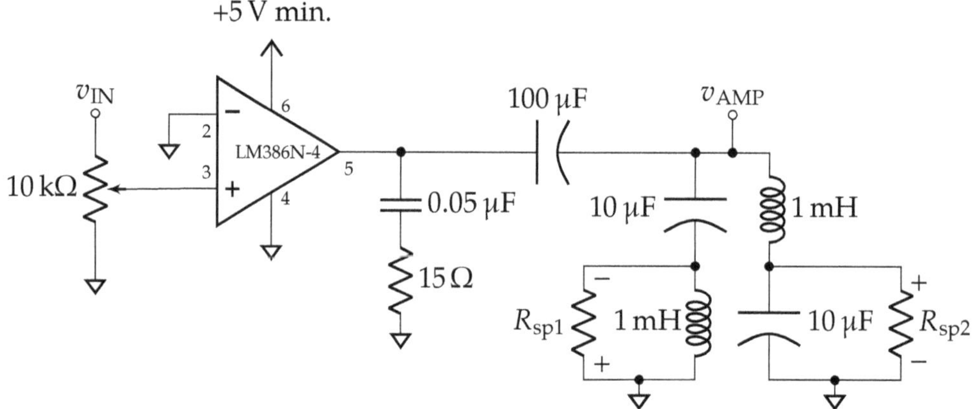

Figure 13.16: Crossover circuit with passive filters

This circuit uses the LM386N-4 low voltage power audio amplifier. You can think think of this circuit as an op amp with a built-in feedback network set to a gain of 20. The feedback is such that the output is biased to one half of the supply voltage, allowing the double-sided input signal to be amplified without clipping on the single-sided supply. The large series capacitor passes only the AC portion of the amplified signal to the crossovers and speakers, returning to a double-sided system.

Driving such a small impedance (that is, large current) is difficult to do while maintaining amplifier performance. The LM386N-4 datasheet [1] specificies that an RC filter on the output is needed in order to operate properly.[a] This circuit also **requires** the use of decoupling capacitors between V_{DD} and ground.

1. What is the role of the potentiometer in this circuit?
2. *Look up* the input resistance of the LM386N-4. Do you expect there to be a problem with loading?
3. *Compute* the expected −3 dB points of each crossover filter, given the resistance of the respective speakers? Analyze the circuit using v_{AMP} as your input and the voltage across R_{sp1} or R_{sp2} as your outputs. You may use a numeric equation solver to solve the equations. Assume that the pass-band gain is 0 dB, so you only need to solve for $\frac{v_{sp,n}}{v_{AMP}} = \frac{1}{\sqrt{2}}$.
4. *Construct* the circuit using the tweeter speaker for R_{sp1} and a woofer for R_{sp2}. The high pass filter adds a 180° phase shift, so it should be hooked up backwards as shown in the schematic.[b]

5. *Measure* the −3 dB points of the crossovers when the speakers are connected. Select the input signal to be small enough that the output sound is quiet but still audible.

6. *Test* the circuit by using an audio source for v_{IN} and listening to the outputs from the speakers. *Describe* how each speaker sounds.

7. *Adjust* the potentiometer and *describe* what it does.

[a] Try removing the 15 Ω resistor after building the circuit and see what happens!
[b] If speakers are not available, 10 Ω resistors can be used as an approximation. Ensure that your are using a small input signal when testing with resistors, as the resistors may get hot and become damaged when using a large signal.

13.9 References

[1] *Lm386 low voltage audio power amplifier*, LM386N-4, SNAS545C, Texas Instruments, May 2017. [Online]. Available: `http://www.ti.com/lit/ds/symlink/lm386.pdf`.

EXPERIMENT 14

Final project: Basic audio equalizer

Due date: This experiment is to be completed over *two* lab sessions. The prelab is due by the *start of the first lab* and the final demonstration and measurements are *due by the end of the second lab.*

Figure 14.1: A multi-band graphical equalizer used for tuning the frequency content of an audio signal. It is made up of many carefully tuned adjustable band-pass filters over the audio frequency range.

14.1 Analog audio amplifier

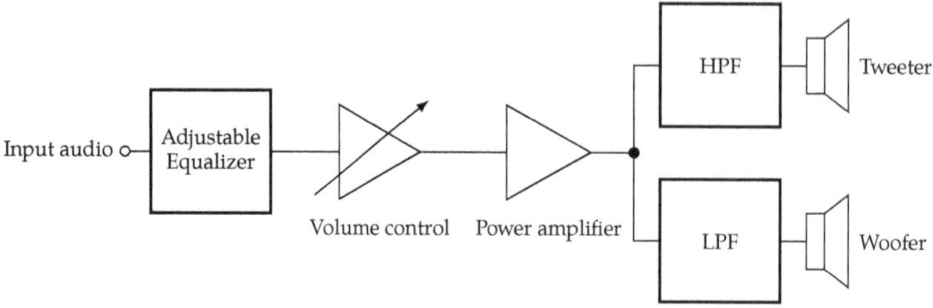

Figure 14.2: Basic audio amplifier block diagram.

An active speaker with a tweeter and woofer will generally contain at least an equalizer, volume control, power amplifier, and crossover subsystems, as shown in figure 14.2. The equalizer adjusts the frequency content of the audio to the listener's preference, the volume control sets the overall loudness, the power amplifier delivers the final signal to the low impedance speakers, and the crossover splits the audio's energy across the low and high frequency speakers. In general, tweeters excel at higher frequency audio and woofers excel at lower frequencies. In this project you will focus on the design of the equalizer circuit and volume control.

14.1.1 Equalizer sub-system

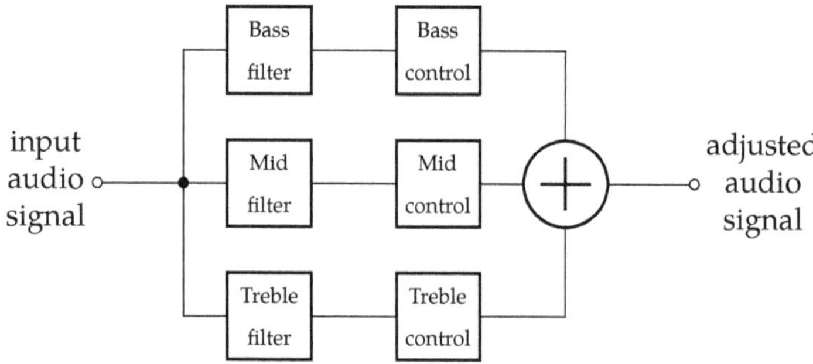

Figure 14.3: The audio equalizer sub-system block diagram.

Equalizers are found in many audio systems, such as stereos and professional audio equipment. The purpose is to adjust the individual loudness of various frequency bands such that the overall sound is to the liking of the listener. They range from simple three frequency band control to 30 or more bands in professional audio equipment. A three band equalizer can be built using the configuration shown in figure 14.3. It contains three filters with adjustable gain (or attenuation).

Figure 14.4: A small audio mixer with a three band equalizer for each channel.

The filters can increase or decrease the amount of bass, mids, or treble in an audio signal based on the position of tuning knobs. Some more advanced equalizers may even let you adjust the frequency and Q for the filters.

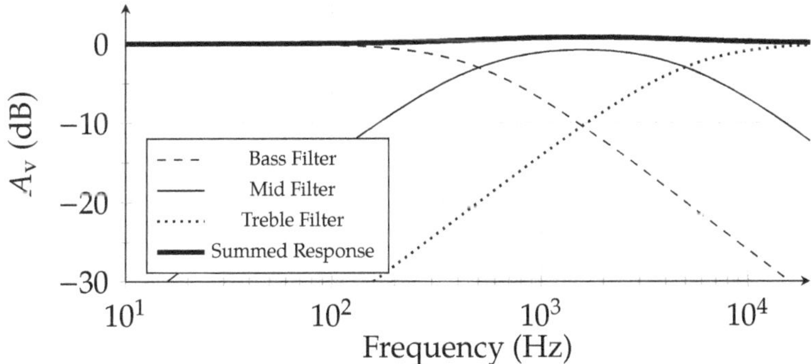

Figure 14.5: Example equalizer frequency response with controls set to maximum.

A three band equalizer sets the relative power of three common bands: bass, mid, and treble. Each band has a filter to isolate the frequencies and a control knob to set the amplification or attenuation. The outputs the filters are summed together to recombine all of the frequency bands. When all of the knobs are adjusted to the same level, a summed frequency response should be relatively flat like in figure 14.5. The low-pass, band-pass, and high-pass filters all split the signal then add it back together, approximately unchanged.

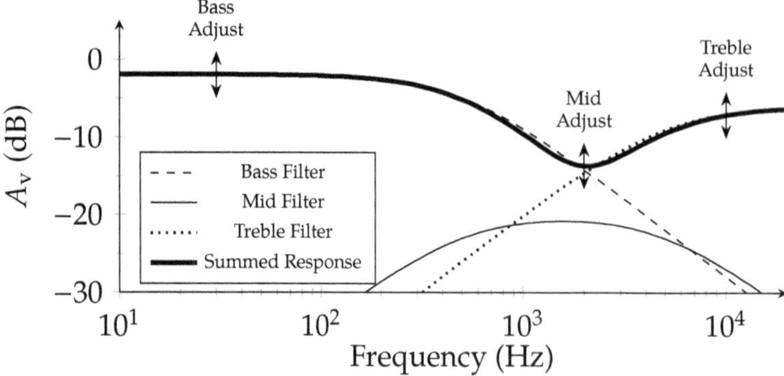

Figure 14.6: Example equalizer frequency response that would provide a bass-boost and mid-cut frequency response.

When the equalizer knobs are adjusted, the frequency response is customized. For example, the frequency response in figure 14.6 has a net effect of increased bass and decreased mids. The equalizer settings can be used to tune the audio system to sound correct based on the recording method, speaker performance, or simply user preference.

Equalizers are implemented in software or hardware. This project focuses on analog equalizers using RC filters and op amp circuits. For the filters, we need to decide what the definition of "bass," "mids," and "treble" are. There is not a globally accepted industry standard, so for the purposes of this project, we will use the following definitions:

Bass The low frequency portion of an audio signal, or everything up to 320 Hz.

Mid The mid frequency portion of an audio signal, or 0.32 kHz to 3.2 kHz.

Treble The high frequency portion of an audio signal, or everything after 3.2 kHz. In practice, the treble is limited to a frequency around the top end of human hearing, in this case we will consider 20 kHz to be highest frequency of concern.

14.1.2 Volume control

The volume control adjusts the overall loudness of the audio signal after each band is scaled and recombined by the equalizer. The volume control subsystem should ensure that output of the equalizer never exceeds the input maximum of the power amplifier. If the output of the equalizer is large enough, the volume control only needs to attenuate the signal; however, it is important to ensure that the volume control is able to function correctly when connected to the amplifier circuit.

14.1.3 Power amplifier

A power amplifier is a circuit that takes a low-power signal and amplifies it to the level needed to drive a given load. Within audio systems, the most common speaker impedances are $4\,\Omega$, $8\,\Omega$, or $16\,\Omega$; so the power amplifier must be able to provide sufficient voltage and current for those loads. A simple op amp is not adequate to drive such small loads, so a purpose-built integrated circuit (IC) is used instead.

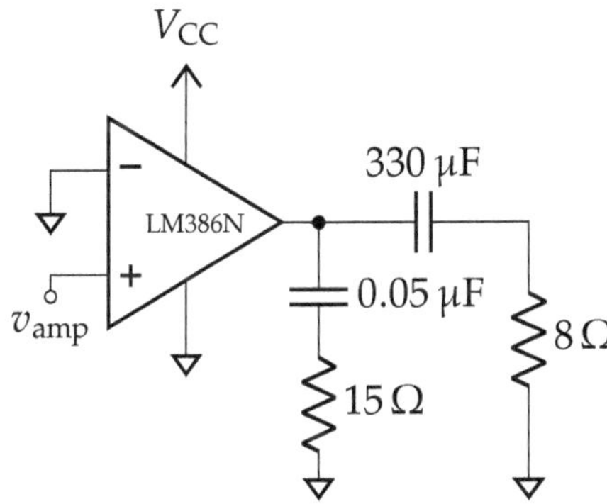

Figure 14.7: Simple LM386N-4 audio amplifier

The low voltage audio power amplifier (LM386N-4) [1] used in experiment 13 is adequate to meet the specifications of this project, but there are a wide variety of amplifier ICs available that can output much more power with significantly less distortion. The circuit in figure 14.7 has a gain of 20 and can drive a speaker as low as $4\,\Omega$. The simplicity of the LM386N-4 makes it a great choice for simple amplifiers, but it lacks many features of higher quality audio amplifiers.

In order to pick V_{CC}, the peak to peak output voltage needs to be calculated from the power specifications and load impedance. The datasheet provides a chart showing the relationship between maximum peak to peak output voltage, load resistance, and supply voltage. If the output voltage is not high enough, there will be significant amounts of distortion, so the audio quality will degrade.(1)

(1) Distortion can be used on purpose for effects, but it isn't so great when you're trying to reproduce existing music well.

14.1.4 Crossover

A crossover is a set of filters that split the high power output so that only energy in the frequency range that a speaker is good at producing is delivered to it. The filters should overlap such that the overall frequency response is flat and the equivalent impedance is as constant as possible across the audio band.

One simple way of achieving this consistency is with a second-order filter called a "Linkwitz Riley" crossover. This filter is designed to have a perfectly flat response when the low-pass response is added to the high-pass response. The filter circuits are shown in figure 14.8.

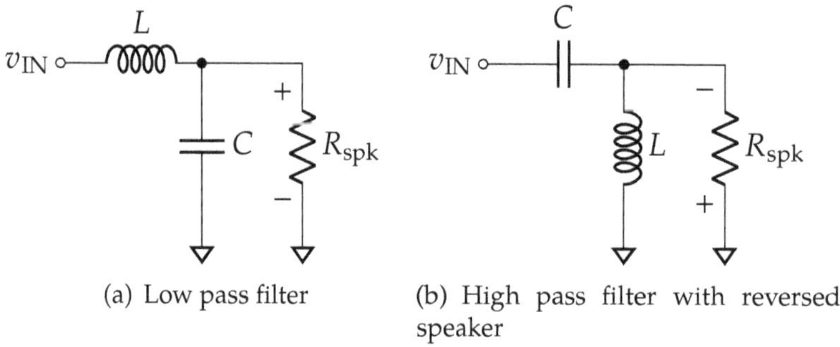

(a) Low pass filter

(b) High pass filter with reversed speaker

Figure 14.8: Possible crossover filters. R_{spk} is the speaker.

The high-pass and low-pass versions of the circuit have the same values with the position of the inductor and capacitor reversed. One thing to note is that the high-pass version of the circuit has a 180° phase shift with respect to the low-pass version, so the tweeter must be hooked up in reverse! Given a speaker resistance and desired cutoff frequency, the inductor and capacitor values can be calculated as follows:(2)

$$L = \frac{R_{\text{spk}}}{\pi f_c}$$
$$C = \frac{1}{4\pi f_c R_{\text{spk}}}$$

(2) These values are calculated by solving for the transfer function of the circuit and finding values that satisfy $\omega_0 = \frac{1}{\sqrt{LC}} = 2\pi f_c$ and $Q = RC\omega_0 = \frac{1}{2}$. This value of Q aligns the cutoff frequency at the −6 dB point, which allows for the two frequency responses to sum together perfectly.

14.2 Specifications

Conditions: All specifications are given assuming volume control is set to maximum with an input 1 $V_{\text{p-p}}$ sine wave. v_{amp} is defined as the voltage at the *input* to the amplifier.

Specification	Requirement
Speaker resistance	8 Ω
Bass filter −3 dB cutoff	320 Hz ±10%
Mid filter −3 dB bandwidth	0.32 kHz to 3.200 kHz ±10%
Treble filter −3 dB cutoff	3.2 kHz ±10%
v_{amp} with all equalizer knobs turned to minimum settings	$< 15\,\text{mV}_{\text{RMS}}$ @ 200 Hz, 2 kHz, and 10 kHz
v_{amp} with all equalizer knobs at maximum settings	$100\,\text{mV}_{\text{RMS}} \pm 10\%$ @ 200 Hz, 2 kHz, and 10 kHz
$v_{\text{amp,max}} - v_{\text{amp,min}}$ max ripple with equalizer at max	$15\,\text{mV}_{\text{RMS}}$ from 200 Hz to 10 kHz
Amplifier output power	> 400 mW from 200 Hz to 10 kHz

14.3 Design of an analog audio equalizer

Your task is to design an analog audio equalizer and volume control, then integrate it with an amplifier and speaker circuit in order to meet the specifications in section 14.2. You should reference the following experiments for circuit inspiration:

1. Root-mean-square (RMS) calculations: experiment 2
2. Voltage dividers: experiment 3
3. Op amp / amplifier circuits: experiment 5, experiment 8, and experiment 13
4. RLC filters: experiment 13

14.4 Prelab

Task 14.4.1: Prelab questions

1. *Design* a circuit topology for the adjustable equalizer and volume control subsystems that should, in principal, meet the project specifications.
2. *Calculate* component values for any filters included in your design.

14.5 Tasks

Task 14.5.1: Tasks

1. *Agree* on a circuit topology within your group. You are encouraged to get feedback on your design from your instructors.
2. *Calculate* the required component values and *verify* those components are available to you. Some components not in your kit are available for you to use.
3. *Compute* the supply voltage necessary for the amplifier:
   a) What is the gain of the amplifier?
   b) What is the output voltage peak to peak when the input is $100\,\mathrm{mV_{RMS}}$?
   c) What supply voltage does the datasheet say will support this output range with an $8\,\Omega$ load?
4. *Construct* and *test* the amplifier and volume control subsystems.
5. *Construct* and *test* each filter independently.
6. *Make* the following measurements
   a) Frequency responses of all filters.
   b) A plot or plots that demonstrate the adder.[a]
   c) A plot or plots that demonstrate low, medium, and high total volume.
   d) A plot or plots that demonstrates that the power amplifier functions correctly.
7. *Integrate* all of the subsystems.
8. *Generate* plots, tables, and/or figures that prove your design meets all of the specifications given in section 14.2.
9. Using your computer as an audio source, *demonstrate* your subsystem to your instructor. Make sure to have all of your prior measurements and your full schematic available for review during the demonstration.

[a] You are tasked with designing an acceptable plot.

Task 14.5.2: Extra credit task

1. *Research* the performance specifications for the provided tweeter and woofer.
2. *Select* a crossover frequency that will pass low frequencies to each speaker within their range of operation.
3. *Design* a crossover circuit to integrate a tweeter and woofer with a crossover frequency of your choosing.
4. *Construct* the crossover integrated with the amplifier output.
5. *Measure* the frequency response of the amplifier plus crossover system.

APPENDIX **A**

Documentation and Reports

One often overlooked skill of an engineer is the ability to document and report work well. A well documented experiment or design is instrumental to making the work easy to use by a customer or other engineer. It is critical that the readers can quickly understand what you have done. Within a single piece of work, an engineer makes many predictions and measurements, all of which have to be reported in a scientifically correct and consistent manner. This is especially true when communicating with other engineers, managers, and customers.

It is important to continuously document the work you are doing.(1) We recommend maintaining a backed-up electronic notebook where you record the steps you are taking and measurements you make. Often, paper records are more convenient while running an experiment, but you should still scan or otherwise digitize your work.

(1) All the authors have lost work and time to inadequate records so we write from experience. Understand that good record keeping requires practice. It is typical not to know what is important at the beginning of a project so thorough records are critical even if you are not sure of the value.

A.1 Formatting fundamentals

Clear and consistent formatting of a report is more than just a method to guide your audience through a report. It makes your report appear more professional and will retain the attention of a reader longer. Good document organization makes reading and understanding the report much easier.(2)

(2) It also makes your grader, boss, or reviewer happier!

The first key piece of organization is the header. The specific header contents and style varies by where your report will be published or shared. For a simple lab report consider placing your name, your collaborators' names, course name, the experiment number/name, and the date at the top of every page.

It can be useful to list the equipment used in the work somewhere after the abstract and before the content. The equipment list helps others reproduce your results and can help identify sources of error. Page numbers in the footer or header are also strongly recommended so that readers can reference specific portions of a report. Page numbers also help with the organization of printed reports.

Every section of a report should have a heading. The easiest solution is to use the premade formatting options in your text editor. As you progress into sections and subsections, consider using progressively smaller headings to show hierarchy.(3) Numbering sections and subsections can help your reader follow the structure of the report and assists in creating a table of contents if the report becomes large.

(3) But don't make them too small!

Finally, we recommend that you pick a standard citation format and use it to cite any references used during the experiment and report writing. The IEEE citation style [1] is easy to use and is recommended for electrical or computer engineering related technical reports.

A.1.1 Signal naming convention

It is important to establish and consistently use a convention for labeling signals. Incorrect or inconsistent usage can make a good report very difficult to understand. This book will use the convention in table A.1. You should consider using the same convention in your reports.

Signal	Definition
X_{Y}	Value of the constant component of X at node Y
x_{y}	Instantaneous value of the varying component of X at time t at node Y
x_{Y}	Sum of the instantaneous values of the varying component of X at time t and the constant part of X at node Y, $X_{\mathrm{Y}} + x_{\mathrm{y}}$
$\mathbf{X}_{\mathrm{Y}}$	Phasor form of a signal X at node Y
Example	v_{OUT}
V_{OUT}	Constant value of the DC component
v_{out}	Instantaneous value of the AC component at time t
v_{OUT}	$V_{\mathrm{OUT}} + v_{\mathrm{out}}$
$\mathbf{V}_{\mathrm{OUT}}$	Phasor form of v_{out}

Table A.1: Signal naming convention

An example of the signal naming convention used in this book is shown in figures A.1 to A.2 The example is a voltage (V) at a node in a circuit called "out," hence V_{OUT}. To express different aspects of the signal, different capitalization is used.

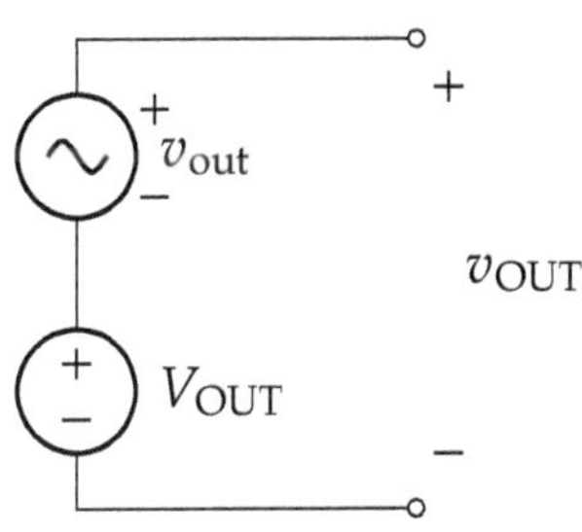

Figure A.1: Example circuit demonstrating the signal naming convention of the book.

v_{OUT} is the instantaneous voltage at node "out" at time t. It consists of a DC component, V_{OUT}, constant for all t summed with an AC component, v_{out}, sampled at time t. The specifics of what constitutes AC or DC will depend on the context of the circuit in use, but you will develop intuition over time as to how to treat certain signals.

A.1.2 Reporting results

To be effective, a measurement must have enough supporting information to be meaningful. In general, measurements should be accompanied by a description of how it was measured (your procedure) as well as the measurements accuracy.

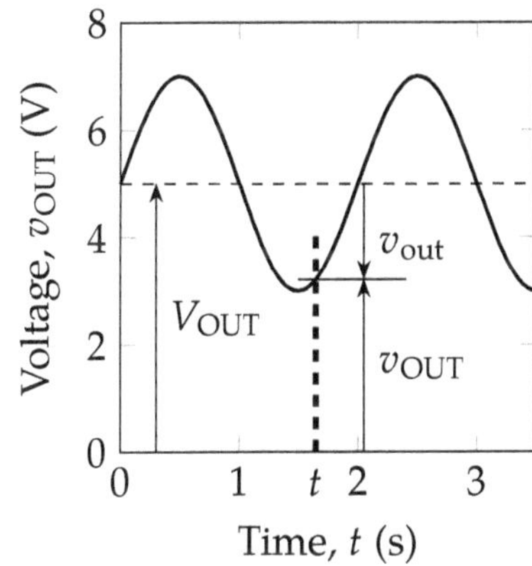

Figure A.2: Example plot demonstrating the signal naming convention of the book from the circuit figure A.1.

Basic reporting of measurements consists of the following:

$$\text{Signal} = \text{Value Unit}$$

or

$$\text{Signal} = \text{Value Unit Accuracy}$$

Accuracy can be expressed in a number of different ways, either absolute or relative to the original value. If the accuracy term is not shown, then it is implied by the number of significant figures (SF) you used to express the value.

It is very important to distinguish between design values that are "perfect" versus measured values from real life. This is covered in greater detail in example A.1.1.

Example A.1.1: Measurements from a potential divider circuit to reduce 10 V to 1 V

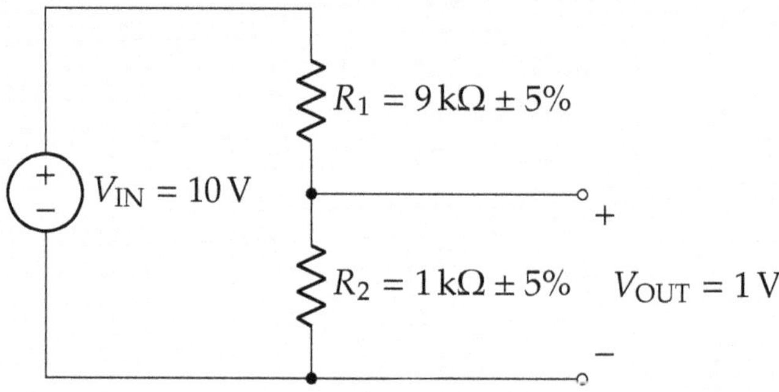

Figure A.3: Circuit diagram of a potential divider with "perfect" design values shown.

In the circuit shown in figure A.3, various design values are shown. Various examples of measurements from the circuit that would be recorded in a design journal, scientific paper, or report are shown in table A.2.

Note in document	Interpretation
Measured $R_1 = 9120\,\Omega \pm 1\%$	The identifier, R_1, is unique and specific; so no mistake is possible. This measurement has an accuracy shown. It is complete.
Measured $R_1 = 9120\,\Omega$	No accuracy measurement posted. Assume at least three SF. The step size at three SF is $10\,\Omega$, so the worst-case accuracy is $^{10}/_{9120} \times 100 = 0.11\%$.
Measured $V_{OUT} = 1.00\,\mathrm{V}$	V_{OUT} shown on the diagram clearly shows the polarity and nodes the voltage is measured across. Three SF are used, so the step is $0.01\,\mathrm{V}$ and the accuracy is $^{0.01}/_{1.00} \times 100 = 1\%$.

Table A.2: Example measurements from the circuit shown in figure A.3.

A.2 Presenting plots

A plot is a graphical way of displaying *measurements* (appendix A.1.2) that describes two (or more) dimensional data. Just as with any simple measurement, it is essential to be specific about what you are plotting.

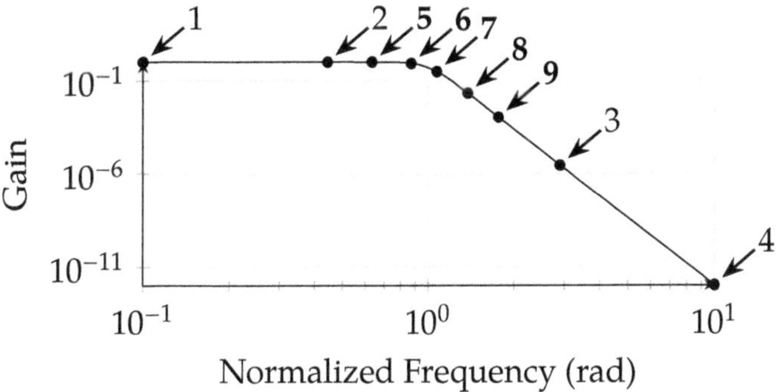

Figure A.4: Example plot constructed from numbered samples. Note the second group of samples, 5–9, provides extra detail to the curve between samples 2 and 3.

For a plot to be complete, be sure to follow the following steps.

1. Include a numbered caption like the captions accompanying every figure in this manual. This is used to refer to the plot from the report text. Be specific so the reader does not get confused.
2. Label axes. This should include the unit of measurement as well as the signal being measured. This is often indicated by a label on a circuit diagram.
3. Include sufficient data points to show important features. For example, refer to figure A.4. The numbered samples are shown with dots. From the numbers you can see samples 1–4 established the range but clearly something interesting happened between samples 2 and 3. To clarify the shape of the line, samples 5–9 where measured in order to to clearly show the "knee" there.
4. Mark all measured data points on the plot or provide a table of data. A continuous line by itself implies that the experimenter sampled every possible point on the line.

When manually sweeping a signal like a voltage, it is common to sweep back and forth several times without recording any data to get a "feel" for the plot. It is then easy to identify ranges that require more samples versus those that require fewer samples.

A.2.1 Circuit diagrams

Circuit diagrams are a standardized way of communicating circuit designs. They are also used to communicate any other aspects that affect electronics through annotations on the schematic. For example, a schematic may indicate which device a decoupling capacitor should be placed close to. The general rules for a circuit schematic are as follows:

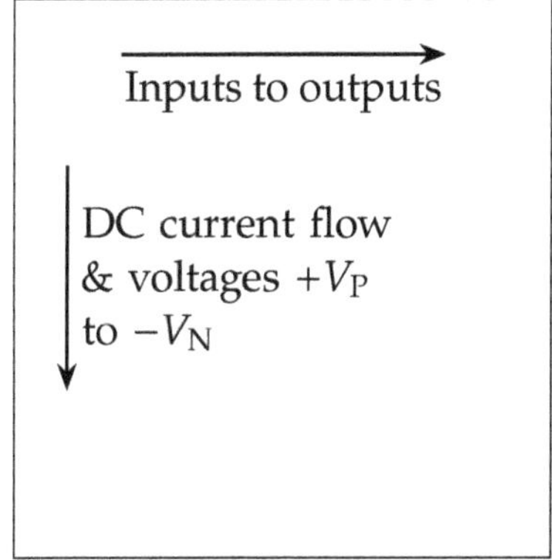

Figure A.5: Guide to circuit diagram layout.

1. DC current flows from top to bottom, which means voltages are arranged with positive voltage supplies at the top and negative voltages or ground at the bottom.

2. Signals flow from left to right, like text in English. So inputs on the left and outputs on the right.

3. All components must use consistent symbols for every device. There are many established conventions for symbols in IEEE Standard 315-1975 [2] and IEEE Standard 91a-1994 [3]. Examples of common symbols we will use are in figure A.6.

4. All devices must be labeled with their values unless the value is unknown.

5. Components should also have a "reference designator" in addition to the value. A reference designator is a letter plus a number that uniquely identifies a part within a schematic. The specific letter to use depends on the part. A listing of common designators is in table A.3.

6. All inputs, outputs, and important nodes should be labeled. Voltages are always measured between two nodes whereas currents are measured through a segment of wire or a two-terminal device.

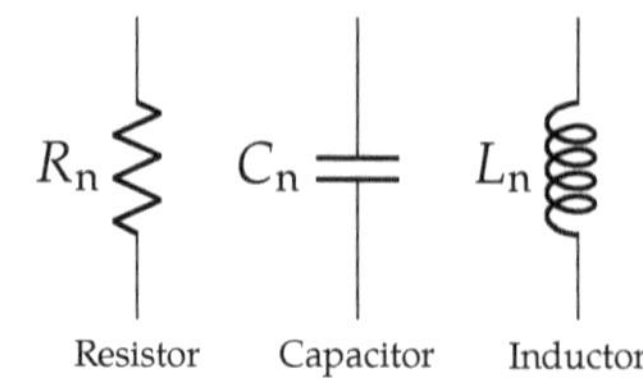

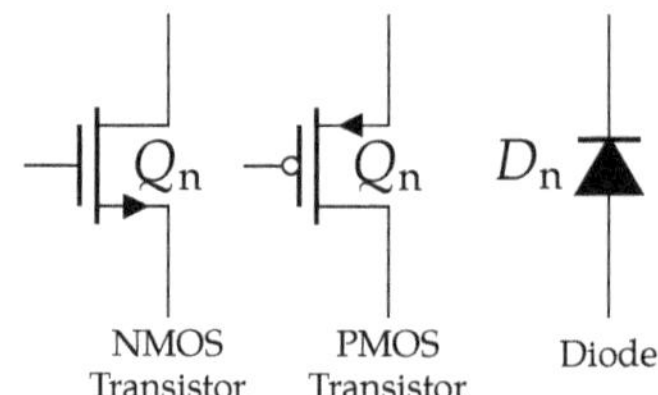

Figure A.6: Common circuit symbols

Do not invent new symbols or designator letters unless absolutely necessary. Circuit schematics are like a language – you cannot simply invent new words for things and expect other engineers to understand you!

Component	Designator Letter
Resistor	R
Capacitor	C
Inductor	L
Transistor(4)	Q, M, N, P
Diode	D
Integrated Circuit	U

Table A.3: Common Reference Designators

(4) Q and M are most common, but occasionally N and P are used to indicate the type of transistor.

A.3 Sharing simulations

A simulation is a method (or technique) for predicting the behavior of a circuit. Simulators are an excellent resource for quickly iterating through design variations and are instrumental to modern circuit design. Simulation program with integrated circuit emphasis (SPICE) tools can be used to simulate almost any circuit. When reporting simulation results, the first key item to report is what simulator you used. This ensures that your simulations can be reproduced by another engineer.

A basic approach to documenting a simulation will do the following:

It is typically acceptable to use the simulation package's default plot formatting for most reports. If you want to make a more professional looking set of plots, then the data can usually be exported as a csv and imported into your preferred plotting tool.

1. Include the circuit design you are simulating. Usually this just involves including a printout of the schematic in the SPICE software. It is incredibly important to ensure that **all** nodes that you will measure are named.

2. Include the SPICE configuration for the test setup. If using LTSpice, this will be printed on the schematic; however, that is not true for all SPICE simulators.

3. Add the resulting plot as a figure in your report. Ensure that all of the waveforms in the plot are clearly labeled and match the names in the schematic.

4. For DC operating point analysis, you can simply report the DC values in the body of the report, but make sure to use the same node names as the simulation schematic. It may sometimes be easier or more illustrative to report the DC measurements as annotations on the schematic.

Example A.3.1: Simulate a 50 Ω, 0 V to 5 V function generator driving a fast switching diode (1N4148) [4] in SPICE

For this example, the software LTspice [5] was used. The schematic in figure A.7 and plot in figure A.8 fully capture the components needed to show a transient simulation.

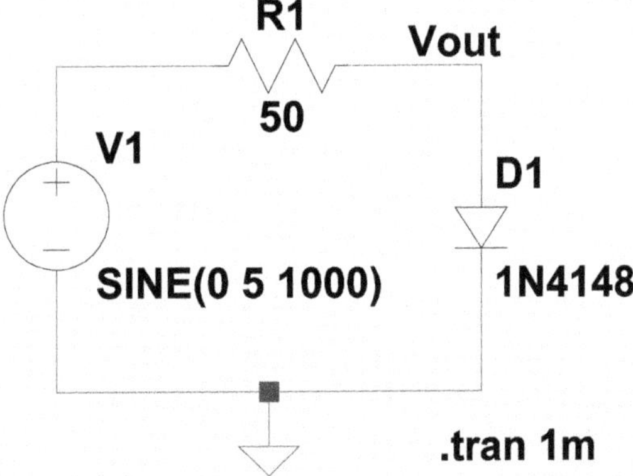

Figure A.7: Circuit diagram of the diode divider drafted in LTspice.

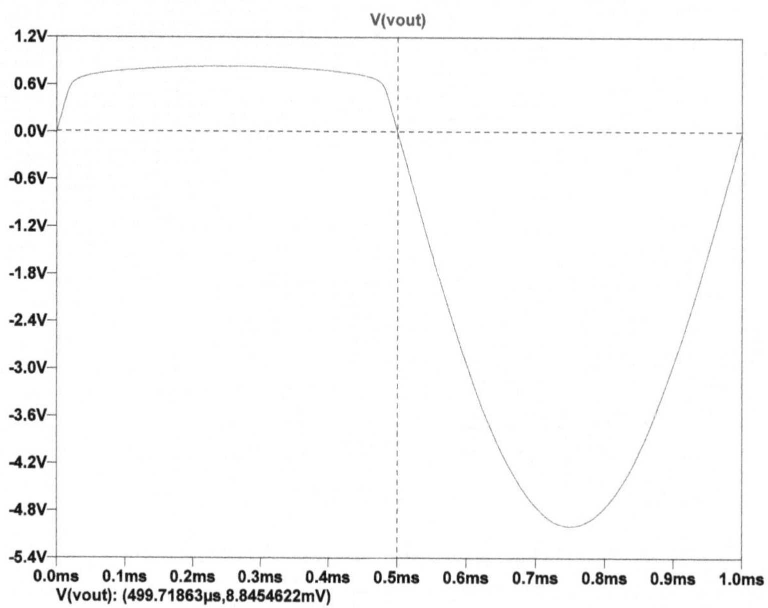

Figure A.8: Transient simulation of the potential divider drafted in LTspice.[a]

[a] LTspice does not follow the signal naming convention of this book – V(vout) should be v_{OUT}.

A.4 References

[1] P. W. Lab, *IEEE Overview.* [Online]. Available: https://owl.purdue.edu/owl/research_and_citation/ieee_style/ieee_overview.html.

[2] *Graphic symbols for electrical and electronics diagrams,* IEEE Std. 315-1975, 1993.

[3] *IEEE standard graphic symbols for logic functions,* IEEE Std. 91a-1991, 1991.

[4] *Small signal fast switching diodes datasheet,* 1N4148, Rev. 1.3, Vishay Intertechnology Inc., Dec. 2013. [Online]. Available: http://www.vishay.com/docs/81857/1n4148.pdf.

[5] A. Devices. (2019). LTspice, [Online]. Available: https://www.analog.com/en/design-center/design-tools-and-calculators/ltspice-simulator.html (visited on 08/01/2019).

APPENDIX B

Glossary

Acronyms

AC alternating, varying with time
ADC analog to digital converter
BJT bipolar junction transistor
CMOS complementary metal–oxide–semiconductor
DC direct, constant in time
DUT device under test
ESR equivalent series resistance
IC integrated circuit
I–V current versus voltage
KCL Kirchhoff's current law
KVL Kirchhoff's voltage law
LED light emitting diode
LTI linear and time invariant
MOSFET metal-oxide-semiconductor field-effect transistor
op amp operational amplifier
PWM pulse width modulation
RLC Resistor-Inductor-Capacitor
RMS root-mean-square
SCPI standard commands for programmable instruments
SF significant figures
SPICE simulation program with integrated circuit emphasis
SR latch set-reset latch
VISA virtual instrument software architecture

Parts

1N4148 fast switching diode
1N5819 schottky barrier diode
FQU17P06 P-Channel QFET® power MOSFET
LDR light dependent resistor
LF356N JFET-input operational amplifier
LM324N Low-Power Quad-Operational Amplifiers
LM339 Quad Differential Comparators
LM386N-4 low voltage audio power amplifier
LMC555 LMC555 CMOS Timer
RFD3055LE N-Channel logic level power MOSFET
UA741 general purpose operational amplifier

APPENDIX C

Labelling Codes of Passive Components

Many different labelling schemes are used to indicate the values of electronic components. We will try to address the most common labelling schemes for through-hole resistors, capacitors, and inductors.

C.1 Resistors

Through hole resistors are labelled using a color code scheme. Most resistors have either 4 or 5 stripes that can be used to determine their value. Resistor codes can be interpreted using table C.1. A 4-band resistor is value-value-multiplier-tolerance, and a 5-band resistor is value-value-value-multiplier-tolerance. Aligning a 4-band resistor is easier because there is a larger gap between the values and tolerance, so it is easy to accidentally read a 5-band resistor backwards. There should be a larger gap between the tolerance and multiplier than between the first two values. When in doubt, check with a DMM.

The resistor in figure C.1(a) has color code yellow-violet-red-gold. This corresponds to $47 \times 100\,\Omega \pm 5\%$, so the resistor is a $4.7\,\text{k}\Omega$ 5% resistor. The second resistor in figure C.1(b) has color code brown-black-black-red-brown, which corresponds to $100 \times 100\,\Omega \pm 1\%$. This resistor is a $10\,\text{k}\Omega$ 1% resistor.

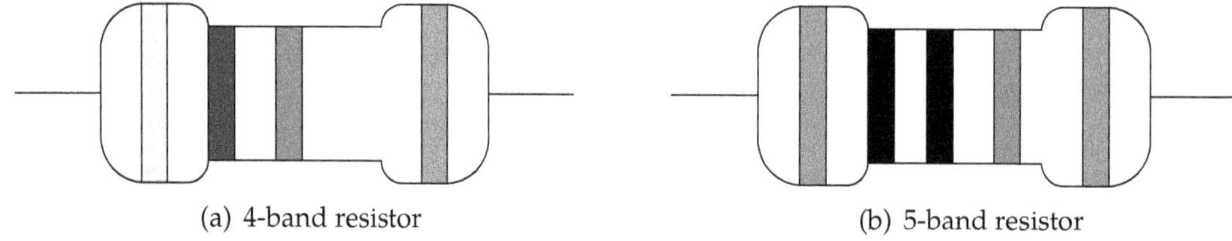

(a) 4-band resistor (b) 5-band resistor

Figure C.1: A $4.7\,\text{k}\Omega$ 5% and $10\,\text{k}\Omega$ 1% resistor

Table C.1: Resistor Color Codes

Color	Value	Multiplier	Tolerance
Black	0	$\times 1\,\Omega$	
Brown	1	$\times 10\,\Omega$	±1%
Red	2	$\times 100\,\Omega$	±2%
Orange	3	$\times 1\,k\Omega$	
Yellow	4	$\times 10\,k\Omega$	
Green	5	$\times 100\,k\Omega$	±0.5%
Blue	6	$\times 1\,M\Omega$	±0.25%
Violet	7	$\times 10\,M\Omega$	±0.1%
Gray	8	$\times 100\,M\Omega$	±0.05%
White	9	$\times 1\,G\Omega$	
Gold		$\times 0.1\,\Omega$	±5%
Silver		$\times 0.01\,\Omega$	±10%

C.2 Capacitors and Inductors

Capacitors and inductors are marked in two ways. The easiest way is with the actual value of the device in the correct units. When this is the case, no table or translation is needed. Unfortunately, the full values don't fit on smaller components, so a different labelling scheme is needed.

Capacitors and inductors are marked with a 3 digit number. The first two digits are the value and the third digit is the power of the multiplier. The resulting number is multiplied by 1 pF for capacitors and 1 µH for inductors. For example, a capacitor marked '474' corresponds to the value 47×10^4. This value is then multiplied by 1 pF, which results in a final capacitance value of 0.47 µF. An inductor marked '103' corresponds to the value 10×10^3, which is then multiplied by 1 µH, so the inductance value is 10 mH.

Capacitors are also usually marked with a tolerance specification, but the labelling strategy varies. The datasheet for the capacitor will always have the most reliable source for interpreting the letters and numbers in the tolerance specification. A common single letter scheme for tolerance is shown in table C.2.

Table C.2: Capacitor Tolerance Codes

Code	Tolerance	Code	Tolerance
B	±0.1%	H	±3%
C	±0.25%	J	±5%
D	±0.5%	K	±10%
F	±1%	M	±20%
G	±2%	Z	+80%/−20%

APPENDIX D

Preferred Number Series for Parts Selection

The "E" series is a "preferred number" range [2] defined by the international standard IEC 60063 [1]. It is a logarithmic scale where the accuracy of the component ensures all the values cover the whole decade range. For example, E24 has 24 values covering the range at ±5% manufacturing tolerance. The E96 has 96 values covering the range at ±1%.

The values of the ranges E24, E12 and, E6 are shown in table D.1.

±5% E24	±10% E12	±20% E6
10	10	10
11		
12	12	
13		
15	15	
16		
18	18	
20		
22	22	22
24		
27	27	
30		
33	33	
36		
39	39	
43		
47	47	47
51		
56	56	
62		
68	68	
75		
82	82	
91		

Table D.1: List of E24, E12 and E6 preferred values for electronics [1].

D.1 References

[1] "Preferred number series for resistors and capacitors," IEC IEC 60063:2015, 2015.

[2] Wikipedia Contributors. (2016). Preferred number — Wikipedia, the free encyclopedia, [Online]. Available: `http://en.wikipedia.org/wiki/Preferred_number` (visited on 07/30/2016).

www.ingramcontent.com/pod-product-compliance
Ingram Content Group UK Ltd.
Pitfield, Milton Keynes, MK11 3LW, UK
UKHW051138260726
13967UKWH00010B/3121